Mohammad-hanif Vora

Study Manual for Farm Machinery and Equipment (Course-I)

Mohammad-hanif Vora

Study Manual for Farm Machinery and Equipment (Course-I)

LAP LAMBERT Academic Publishing

Impressum / Imprint

Bibliografische Information der Deutschen Nationalbibliothek: Die Deutsche Nationalbibliothek verzeichnet diese Publikation in der Deutschen Nationalbibliografie; detaillierte bibliografische Daten sind im Internet über http://dnb.d-nb.de abrufbar.

Alle in diesem Buch genannten Marken und Produktnamen unterliegen warenzeichen-, marken- oder patentrechtlichem Schutz bzw. sind Warenzeichen oder eingetragene Warenzeichen der jeweiligen Inhaber. Die Wiedergabe von Marken, Produktnamen, Gebrauchsnamen, Handelsnamen, Warenbezeichnungen u.s.w. in diesem Werk berechtigt auch ohne besondere Kennzeichnung nicht zu der Annahme, dass solche Namen im Sinne der Warenzeichen- und Markenschutzgesetzgebung als frei zu betrachten wären und daher von jedermann benutzt werden dürften.

Bibliographic information published by the Deutsche Nationalbibliothek: The Deutsche Nationalbibliothek lists this publication in the Deutsche Nationalbibliografie; detailed bibliographic data are available in the Internet at http://dnb.d-nb.de.

Any brand names and product names mentioned in this book are subject to trademark, brand or patent protection and are trademarks or registered trademarks of their respective holders. The use of brand names, product names, common names, trade names, product descriptions etc. even without a particular marking in this work is in no way to be construed to mean that such names may be regarded as unrestricted in respect of trademark and brand protection legislation and could thus be used by anyone.

Coverbild / Cover image: www.ingimage.com

Verlag / Publisher:
LAP LAMBERT Academic Publishing
ist ein Imprint der / is a trademark of
OmniScriptum GmbH & Co. KG
Bahnhofstraße 28, 66111 Saarbrücken, Deutschland / Germany
Email: info@lap-publishing.com

Herstellung: siehe letzte Seite /
Printed at: see last page
ISBN: 978-3-659-81125-8

STUDY MANUAL FOR FARM MACHINERY AND EQUIPMENT
(Course 1)

Section I (Practical) & Section II (Theory)

Edited by

M. D. Vora

PREFACE / ACKNOWLEDGEMENT

The 'Study Manual for Farm Machinery and Equipment-I (One)' is prepared with a view to assist the students pursuing the course on Farm Machinery and Equipment-I under the discipline of Agriculture and Agricultural Engineering which consists of exercise and study materials on farm machinery and equipment related with different initial farm operations which include earth moving, seedbed preparation (primary and secondary tillage), sowing, weeding, plant protection (spraying and dusting) etc. The related aspects as contained under the drawn syllabus at national level are attempted to be covered at best. However, the readers are requested to treat this publication as a part of the study process and not as a final output. Corrective suggestions for improvement of the contents are very welcome. The author expresses his gratitude to all contributors whose literature has been referred in preparation of this booklet all of whom are not substantially acknowledged at all places.

- M D Vora (Assistant Professor, Farm Machinery and Power Engineering, CAET, AAU, Godhra, Gujarat, PIN Code 389001)
mdvora11@gmail.com
mdvora@aau.in

CONTENTS

Section-I

PRACTICAL 1

1. TITLE OF PRACTICAL :
   INTRODUCTION TO VARIOUS MACHINES AND IMPLEMENTS AVAILABLE IN FARM MACHINERY LABORATORY

2. AIMS / OBJECTIVES :
   (i)   To prepare a list of various farm machinery and implements available in the farm machinery laboratory
   (ii)   To classify the various farm machinery and implements according to the type of field operations to be conducted by them
   (iii) Write the functional characteristics of different farm machines and implements

3. DESCRIPTION OF PRACTICAL :
   The farm machinery laboratory in the Department of Farm Machinery and Power Engineering at CAET-Godhra was visited to observe the various farm machinery and implements available in the laboratory. After preparing a list of the various farm machinery and implements available in the lab, they were classified into the type of implements and machines as per their usage for conducting various field operations on the agricultural farm.

   The farm machinery and implements are broadly classified into following categories:
   (a) Farm implements for seed-bed preparation
   (b) Farm implements for conducting the sowing or seeding operations
   (c) Weeding implements and machines
   (d) Plant protection equipment
   (e) Harvesting and threshing equipment
   (f) Earth moving equipment

   The functional characteristics are described in the following table:
   Table 1.1: Different kind of farm implement / machine and its usage

Sr. no.	Name of farm implement / machine	Use of farm implement / machine
1		
2		
3		
4		
5		
6		
7		

PRACTICAL 2
1. TITLE OF PRACTICAL :
   CONSTRUCTION, ADJUSTMENT AND WORKING OF PRIMARY TILLAGE IMPLEMENTS

2. AIMS / OBJECTIVES :
   (i)   To study the construction, adjustment and working of Desi (Indigenous) Plough

[1]

(ii)   To study the construction, adjustment and working of M.B. Plough

(iii)   To study the construction, adjustment and working of Disc Plough

3. IMPLEMENTS REQUIRED :
   (i)   Desi Plough (Indigenous Plough)
   (ii)   Mould Board Plough
   (iii)   Disc Plough

4. METHOD :

Constructional details of the above said implements were recorded. Various adjustments which can be made in above ploughs are studied. Working of the ploughs was observed.

5. OBSERVATION :

5.1   Construction of Desi Plough is studied by observing its various components in the farm machinery laboratory. Materials of construction of different parts were studied.

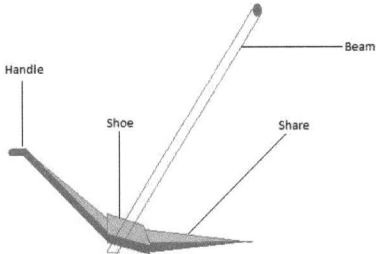

Fig. 2.1: Construction of Desi (Indigenous) plough

(a) Constructional details of Desi Plough

Table 2.1: Parts of Desi (Indigenous) Plough

Sr. no.	Name of the part	Materials of construction

(b) Adjustment of Desi Plough

Controlling the depth of ploughing by varying the angle made working tool with the ground surface (varying the line of pull with the direction of travel)

(c) Working of Desi Plough

Furrow cut by the Desi Plough was triangular in shape.

5.2   Construction of Mould Board Plough is studied by observing its various components in the farm machinery laboratory. Materials of construction of different parts were studied.

(a) Constructional details of M.B. Plough

Table 2.2: Parts of M.B. (Mould Board) Plough

Sr. no.	Name of the part	Materials of construction

[2]

(b) <u>Adjustment of M.B. Plough</u>
Horizontal suction and Vertical suction

(c) <u>Working of M.B. Plough</u>
Furrow cut by the M.B. Plough was trapezoidal in shape. Soil is better pulverized and inverted in comparison to the Desi Plough.

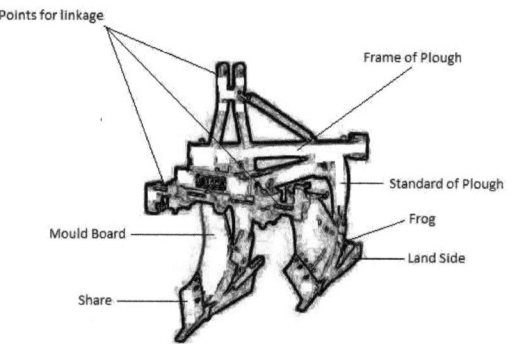

Fig. 2.2: Mould-board plough – components of construction

5.3   Construction of Disc Plough is studied by observing its various components in the farm machinery laboratory. Materials of construction of different parts were studied.

(a) <u>Constructional details of Disc Plough</u>
Table 2.3: Parts of Disc Plough

Sr. no.	Name of the part	Materials of construction

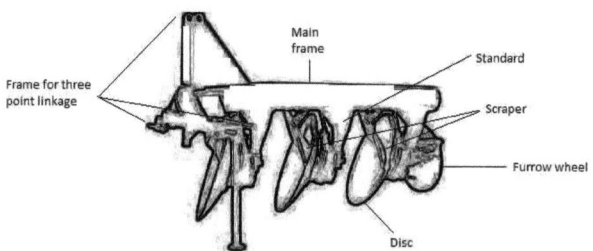

Fig. 2.3: Disc Plough – Parts of its construction

[3]

(b) Adjustment in Disc Plough
Disc angle and Tilt angle

(c) Working of Disc Plough
The sharp edges of the discs facilitate the easy penetration of the discs into the soil. Soil is better turned and trashes are removed by the provision of scrapers on the discs of the plough. Disc plough is suitable for stony soils.

PRACTICAL 3

1. TITLE OF PRACTICAL :
   CONSTRUCTION, ADJUSTMENT AND WORKING OF SECONDARY TILLAGE IMPLEMENTS

2. AIMS / OBJECTIVES :
   (i)   To classify different types of harrows used for conducting secondary tillage operations
   (ii)   To study the construction, adjustment and working of Disc harrows
   (iii)   To study the construction, adjustment and working of Drag harrows

3. DESCRIPTION OF PRACTICAL :
   Secondary tillage operations
   Secondary tillage operations are those operations which are conducted after primary tillage operations. Secondary tillage operations comprise the breaking of clods, pulverizing the ploughed soil, cutting the crop residues and mix with soil, leveling the field etc.

   Secondary tillage operations may be broadly classified into following categories:
   (i)   Disc harrows
   (ii)   Drag harrows

   Disc harrows
           Disc harrows are made uniformly spaced steel discs mounted on gang(s). In case of disc harrows made of two gangs, the steel discs are gang wise mounted in opposite directions. Hence when operated into the field, the right side gang will throw the soil towards right and left side gang will throw the soil towards left.

   Drag harrows
           Drag harrows are used since long than the disc harrows. These harrows are used to crush the clods, to break the soil crust, to level the ground and sometimes to cover the seeds.

4. MATERIALS / INSTRUMENTS / EQUIPMENT REQUIRED :
   All secondary tillage implements available in the farm machinery laboratory of the FMPE department.

5. METHOD :
   Secondary tillage implements available in the farm machinery laboratory of the department of FMPE are used for the study.

6. OBSERVATIONS AND DISCUSSION:
   Following harrows were studied:

[4]

    (i)   Disc harrow
    (ii)   Spike tooth harrows
    (iii)   Spring tyne harrows
    (iv)   Blade harrows
    (v)   Cultivators

Disc harrows

Disc harrows may be classified as below:

(a) On basis of source of power, disc harrows may be animal drawn or tractor operated
(b) On basis of the number and configuration of the gangs, the disc harrows may be classified into:
    (i) single action disc harrow
    (ii) double action disc harrow
    (iii) tandem type disc harrow
    (iv) offset type disc harrow

The single action disc harrow consists of one gang with discs mounted on it or it may consists of two gangs arranged besides each other with steel discs mounted in opposite directions on each gang i.e. right side gang throws soil towards right and left side gang throws soil towards left. In case of double acting disc harrows, the one gang is followed by another gang working in opposite direction. Tandem type disc harrow contains four gangs with four sets of gangs. Front placed two gangs throw the soil outward and rear gangs throw the soil inwards hence operating in the ground twice. Offset disc harrows are arranged to be drawn behind the tractor in offset position having two gangs, one followed by another, throwing the soils in opposite directions.

Main Components of disc harrow are:

1.
2.
3.
4.
5.
6.
7.
8.

Drag harrows

Some of the drag kind of harrows are listed at below.

1. Spike tooth harrow
2. Spring tooth harrow
3. Knife harrow
4. Patela
5. Triangular harrow
6. Zig-zag harrow
7. Roller harrow, Rotary harrow etc.
8. Clod crusher

Blade harrows

Blade harrows are very common among the farmers. In clayey soils, the blade harrows are used for preparing the seed beds for sowing of the kharif as well as rabi season crops.

Cultivators

Cultivator is widely used by the farmers for conducting the secondary or light tillage operations. Cultivators manipulate the soil with the help of shovels or sweeps attached to its tines or standards. It uproots the weeds grown in the agricultural field. Cultivators are also used for seed bed preparation.

Adjustments in Disc harrow:
Greater penetration by the disc harrow can be obtained by following measures:
- (i) Adding additional weight on the harrow
- (ii) Lowering the hitching point
- (iii) Using sharpened discs
- (iv) Using discs of less concavity
- (v) Decreasing the speed of operation

Types of Cultivators
(a) Trailed type cultivator: It is trailed by hitching arrangement. A pair of wheels is provided.
(b) Mounted cultivator: It is mounted on tractor by three point linkage and operated by hydraulic system of the tractor.
(c) Cultivator with rigid tynes: Tynes are fitted on the main frame by clamps and bolts.
(d) Cultivator with spring loaded tynes: Tynes are hinged to the frame with provision of springs to deflect when any obstacle is faced during its operation.

Working tools of Cultivator
Different types of shovel and sweep are used working tools by fitting them on the cultivator tynes. Several types of shovel and sweep are shown in Fig. --.

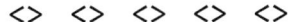

PRACTICAL 4

1. TITLE OF PRACTICAL :
    CONSTRUCTIONAL DETAILS, ADJUSTMENT AND WORKING OF SEED-CUM-FERTILIZER DRILLS AND PLANTER

2. AIMS / OBJECTIVES :
    - (iv) To study different sowing methods for agricultural crop production
    - (v)   To prepare a list of various implements used for sowing operations
    - (vi) To study the constructional features of a seed-cum-fertilizer drills and planters
    - (vii) To study working and adjustment of seed drills and planters

3. DIFFERENT SOWING METHODS :
    Table 4.1: Different methods of sowing

Sr. no.	Types of sowing methods	Description
1	Broadcasting	
2	Dibbling	
3	Seed dropping behind the plough	
4	Hill dropping	
5	Drilling	
6	Check-row sowing	

7	Planting	

4. IMPLEMENTS USED FOR SOWING OPERATION

Table 4.2: Different implements used for sowing of agricultural crops

Sr. no.	Name of farm implement/machine	Description of implement observed in lab./field
1	Dibblers	Naveen Dibbler is observed
2	Seeding behind the plough	A seeding tube with hopper is fitted behind the plough
3	Animal drawn seed drills	Pulled by a pair of animals/bullocks
4	Tractor mounted seed drills	Provided with a three point hitching system
5	Tractor mounted seed-cum-fertilizer drills	Separate boxes for seeds and fertilizer for suitable metering mechanisms
6	Planters	Provided with inclined seed plates
7		

5. CONSTRUCTIONAL FEATURES OF SEED-CUM-FERTILIZER DRILL

Table 4.3: Parts of construction of seed-cum-fertilizer drills and planters

Sr. no.	Name of parts	Materials of construction and usage
1	Frame	
2	Seed box and fertilizer box (hoppers)	
3	Seed metering device	
4	Furrow openers	
5	Seed tubes	
6	Ground wheels	
7	Chain and sprocket mechanism for power transmission	
8	Lever for lifting and lowering of the ground wheel	
9	Transport wheels	Provided for movement of machine from one place to another

6. WORKING AND ADJUSTMENT OF SEED DRILLS/PLANTERS

Difference Between Seed Drill And Planter

PRACTICAL 5
1. TITLE OF PRACTICAL :
   CALIBRATION OF SEED DRILL

2. AIMS / OBJECTIVES :
   (i) To calibrate the seed drill available in the farm machinery lab to identify the present seed rate
   (ii) Adjustment of the seed metering mechanism to obtain the desired seed rate

3. Materials and Equipment required
   (a) Maize seeds, 2 kg
   (b) Seed drill / Seed-cum-fertilizer drill / Planter available in farm machinery lab
   (c) Bowls for collection of seeds
   (d) Weighing balance

4. Description of practical
   Calibration of seed drill
   The procedure of testing the seed drill for getting correct amount of seed rate is called calibration of seed drill

   Purpose of calibration
   Calibration is performed to confirm the drilling of a predetermined seed rate of the seed drill or planter

   Requirement of calibration
   It is advisable to calibrate the seed drill before using in the field for actual sowing

   Method of calibration
   Following steps are followed for conducting the calibration process:
   i) Determine working width of the seed drill.
   w = n x d, Where w = width of seed drill (m), n = number of furrow openers and s = spacing between two furrow openers (m)
   ii) Measure the diameter of ground wheel and calculate the circumference of wheel
   iii) Find the length of strip (L) required to cover $1/25^{th}$ of 1 hectare area
   iv) Calculate the number of revolutions (N) required to travel the distance 'L'
   v) Make a mark on the ground wheel
   vi) Lift the ground wheel (drive wheel) attachment so that wheel can rotate freely
   vii) Place the appropriate and similar quantity of seeds in the seed boxes
   viii) Place the bowls below the furrow openers
   ix) Record the setting of metering for rate control
   x) Revolve the ground wheel at a uniform speed for 'N' revolutions
   xi) Weigh the quantity of seeds collected at the end of 'N' revolutions
   xii) Calculate the seed rate in kg/ha.

5. Observation
   (a) Number of furrow openers of the seed drill under calibration (n): _____
   (b) Spacing between two furrow openers (d): _____
   (c) Width of seed drill (w = n x d): _____
   (d) Length of strip (L) required to cover the $1/25^{th}$ ha i.e. 400 m^2 area (400/w): _____
   (e) Diameter of ground wheel (D): _____
   (f) Circumference of drive wheel (π D):_____
   (g) Number of revolutions of required to rotate the drive wheel to cover the distance 'L'

   (Number of revolutions = $\dfrac{L}{\pi.D}$): _____

[8]

(h) Weight of seeds collected in the bowls: _____ kg
(i) Weight of seeds collected in the bowls after rate control adjustment (changed setting of metering device): _____ kg

Table 5.1: Existing seed rate observed at present setting of metering

Number of revolutions of driving wheel	Weight of seeds (in g)						Total weight of seeds	
	Furrow opener – 1	Furrow opener - 2	Furrow opener - 3	Furrow opener - 4	Furrow opener - 5	Furrow opener - 6	In g	In kg

Table 5.2: New seed rate observed after adjustment in metering

Number of revolutions of driving wheel	Weight of seeds (in g)						Total weight of seeds	
	Furrow opener – 1	Furrow opener - 2	Furrow opener - 3	Furrow opener - 4	Furrow opener - 5	Furrow opener - 6	In g	In kg

6. Calculation
   (a) Existing seed rate in kg/ha observed at present setting: _____

   (b) New seed rate in kg/ha (at changed setting of metering): _____

   Remarks:

7. Conclusion if any.

<> <> <> <> <>

PRACTICAL 6

1. TITLE OF PRACTICAL :
   MEASUREMENT OF FIELD CAPACITY AND FIELD EFFICIENCY OF MOULD-BOARD PLOUGH

2. AIMS / OBJECTIVES :
   (i) To calculate theoretical field capacity of an M.B. Plough
   (ii) To measure effective field capacity of M.B. Plough
   (iii) To calculate field efficiency of M.B. Plough

3. Description of practical
   Performance of farm implements / machines in the field is generally expressed by field capacity and field efficiency.

   Field capacity
   It is the rate of field coverage. Field capacity may be Theoretical and Actual.

[9]

Theoretical field capacity

Theoretical field capacity may be calculated by the following formula:

$$\text{Theoretical field capacity (in ha/hr)} = \frac{w \times s}{10}$$

Where, w = width of cut (in m) and s = speed of operation (in kmph)

Effective field capacity

Effective field capacity may be expressed as below:

$$\text{Effective field capacity (in ha/hr)} = \frac{w \times l}{t \times 10}$$

Where, w = width of cut (in m), l = length of strip (in km) and t = time taken (in hour)

Field efficiency

Field efficiency may be calculated as following:

$$\text{Field efficiency (in \%)} = \frac{\text{Effective field capacity}}{\text{Theoretical field capacity}} \times 100$$

4. Materials and Equipment required
   - (i) Tractor
   - (ii) Mould-board plough
   - (iii) Suitable field for conducting ploughing operation
   - (iv) Stop watch
   - (v) Scale

5. Method
   - Width of M.B. plough may be obtained from multiplying the number of plough bottoms (No.) and spacing between two bottoms (in metre)
   - For calculating the theoretical field capacity (using equation), speed is observed directly from the speedometer during ploughing operation using M.B. plough attached with tractor
   - For measuring effective or actual field capacity, the tractor attached with M.B. plough is operated for a specific length i.e. 100 m and time taken is noted
   - Applying the equation, the actual (effective) field capacity can be calculated
   - Field efficiency may be calculated from the ratio of effective field capacity and theoretical field capacity

6. Observations and Calculation

   Table 6.1: Observation and calculation of theoretical field capacity (ha/hr)

No. of plough bottoms n (No.)	Spacing between two plough bottoms d (m)	Width of plough $w = n \times d$ (m)	Average speed of operation s (kmph)	Theoretical field capacity F.C. (Theo.) $= \frac{w \times s}{10}$ (ha/hr)
(1)	(2)	(3) = (1) × (2)	(4)	(5)

Table 6.2: Observation and calculation of effective field capacity (ha/hr)
(Length of strip, l = 100 m i.e. 0.1 km)

No. of plough bottoms n (No.)	Spacing between two plough bottoms d (m)	Width of plough w = n x d (m)	Time taken to travel 100 m distance in the field by the tractor attached with M.B. plough 't' (hour)	Effective field capacity F.C. (Effe.) = $\dfrac{w \times l}{t \times 10}$ (ha/hr)
(1)	(2)	(3) = (1) x (2)	(4)	(5)

Calculation of field efficiency:

$$\text{Field efficiency (in \%)} = \frac{\text{Effective field capacity}}{\text{Theoretical field capacity}} \times 100$$

Practical Exercise 6.1:
To compute theoretical field capacity, actual field capacity and field efficiency.

Solve:

A tractor is operated for a day of 8 hours at the speed of 6 kmph for ploughing a field using a 2-bottom 45 cm M.B. plough. Calculate the **theoretical field capacity of ploughing**. During initial two hours of working, tractor with M.B. plough traveled around 10.0 km. Calculate **actual field capacity, field efficiency and total area ploughed during a day.**

Solution

Theoretical field capacity of ploughing (in ha/hr) = $\dfrac{w \times s}{10}$

Where w = width in m and s = speed in kmph

Theoretical field capacity of ploughing (in ha/hr) = $\dfrac{w \times s}{10}$

$$= \frac{0.9 \times 6}{10}$$

$$= 0.54 \text{ ha/hr} \quad \ldots \quad \ldots \text{Ans-1.}$$

[11]

Tractor with an M.B. plough has traveled a distance of 39.5 km in 8 hours.

$\therefore$ Actual field capacity

$$= \frac{w \times l}{t \times 10000}$$

$$= \frac{0.9 \times 10.0}{2 \times 10} = 0.45 \text{ ha/hr} \quad \dots \quad \dots \text{Ans-2.}$$

$\therefore$ Field efficiency (in %)

$$= \frac{\text{Actual field capacity}}{\text{Theoretical field capacity}} \times 100$$

$$= \frac{0.45}{0.54} \times 100$$

$$= 83.33\% \quad \dots \quad \dots \quad \dots \text{Ans-3.}$$

Area ploughed during a day can be calculated from actual field capacity, which is 0.45 ha/hr.

$\therefore$ Total area ploughed during a day (i.e. 8 hours) will be

Actual field capacity x 8 = 0.45 x 8

$$= 3.6 \text{ ha} \quad \dots \quad \dots \quad \dots \text{Ans-4.}$$

<> <> <> <> <>

PRACTICAL 7

1. TITLE OF PRACTICAL :
   CONSTRUCTION DETAILS, ADJUSTMENT AND WORKING OF PLANT PROTECTION EQUIPMENT

2. AIMS / OBJECTIVES :
   (i)   To study different types of spraying and dusting equipments available in the laboratory of farm machinery and equipments
   (ii) To study basic components of sprayers
   (iii) To calibrate spraying capacity of a sprayer

3. Materials and Equipment required
   Different sprayers available in farm machinery and equipments laboratory

4. Description
   Table 7.1: List of spraying and dusting equipments available in the Laboratory and their features

Sr. no.	Name of spraying / dusting equipment	Main features
1		
2		
3		
4		
5		
6		
7		
8		

Table 7.2: Basic components of sprayers and their functional details

Sr. no.	Name of component	Functional details
1	Nozzle body	
2	Swirl plate	
3	Spray gun	
4	Spray boom	
5	Filter	

6	Relief valve	
7	Pressure regulator	
8	Cut-off valve	
9	Nozzle disc	
10	Nozzle boss	
11	Spray lance	

5. Calibration of sprayer
The spray volume required to cover the particular area under spraying during a specific time period depends on:

  - Rate of discharge by nozzles
  - Forward speed of travel
  - Width of spraying swath

It is advisable to calibrate the sprayer by having a trial spraying using plain water on a piece of specific area. Based on these, the quantity of spraying solution required to cover the unit area under spray can be worked out using following equation:

Rate of spray (L/ha) = Spray volume (L) on trial plot / Area (ha) of trial plot

*L/ha = Litres per hectare.

<> <> <> <> <>

PRACTICAL 8

1. TITLE OF PRACTICAL :
CONSTRUCTION, ADJUSTMENT AND WORKING OF EARTH MOVING EQUIPMENT

2. AIMS / OBJECTIVES :
   (i)   To study the constructional and functional features of earth moving equipments
   (ii)   To study different types of blades for dozer and their adjustments for conducting various field operations
3. Materials and Equipment required
Earth moving equipments available in farm machinery laboratory such as dozer
4. Description
Table 8.1:   Constructional and/or functional features of various earth-moving equipments

Sr. no.	Name of earth moving equipment	Constructional and/or functional features
1		
2		
3		
4		
5		

Table 8.2: Different types of blades used for mounting on dozers

Type of blade	Constructional feature	Functional feature
Straight blade		

| Angular blade | | |
| Special purpose | | |

Table 8.3: Types of field operations and respective settings of the Dozer blade

Field operation required	Adjustment and working of Dozer blade
To conduct digging into the field	
To transport the material from one place to another	
To spread the material on to the field	
To cut V-ditches of shallower depth	
To create a stock-pile	
To spread the stock-pile	
Backfilling	

PRACTICAL 9

1. TITLE OF PRACTICAL :
   CONSTRUCTION DETAILS, ADJUSTMENT AND WORKING OF ROTAVATOR AND ROTARY TILLERS

2. AIMS / OBJECTIVES :
   (i)   To understand operating principle of rotary tillers
   (ii)   To study main components of rotary tillers
3. Materials and Equipment required
   Rotavator available in farm machinery laboratory
4. Principle of working of rotary tillers
   Rotary tiller is a tillage machine used for preparing the land suitable for seedbed without much over-turning of the soil. It eradicates weeds by rotating sharp edged blades. It applies engine power of the tractor through PTO shaft to obtain the rotary power. Draft requirement is reduced as compared to conventional tillage. Tractor mounted rotary tiller is also commonly known as rotavators. Two or three pairs of blades are used on each flange clamped to the rotor shaft of the rotavator. As the rotor shaft revolves, blades cut the slices from the soils.

5. Construction of rotavator
   Components of rotavators are described at below.

Table 9.1:   Main components of rotary tillers

Sr. no.	Name of component	Constructional/functional features
1	Propeller shaft	
2	Slip clutch	
3	Gear box	
4	Slip drive	
5	Blades	
6		
7		
8		

[14]

6. Power transmission in rotavator

Rotavator receives the power from the power take off (PTO) shaft of the tractor in operation. A shielded double universal joint shaft is used to connect with the gear box of the rotavator. The gear box hosts a crown wheel and pinion, pairs of inter-changeable spur gears and rotor engagement clutch. Multiple plate clutch is provided before power input to the gear box.

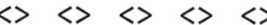

PRACTICAL 10

1. TITLE OF PRACTICAL :
   WORKING OF WEEDING EQUIPMENT
2. AIMS / OBJECTIVES :
   (i) To study manual weeding tools
   (ii) To study animal drawn and tractor operated weeding implements
   (iii) To study power weeder

3. Materials and Equipment required
   Weeding tools and implements available in the farm machinery laboratory

4. Description of weeding tools/implements
   Several hand tools for performing manual weeding operation are discussed at below.
   (a) Khurpis
       Among traditional hand tools used for weeding, khurpis made by local artisans are popular among small and marginal farmers. Khurpis may vary in shape, size and weights. It consists of a cutting blade appropriately fitted with a small wooden handle. A person uses it while sitting down on the heels and moving forth. (See figure)
   (b) Sickles
       Sickle is a very common tool used among the farmers for weeding. Different types of sickles are used on the farms e.g. (i) Plain and Serrated sickles, (ii) Traditional and Naveen sickles
   (c) Hand hoe
       It is a hand tool mostly attached to a wooden stick and used for inter-cultivation (mainly weeding and mulching). Blades or tynes are used as working tools. (See figure)
   (d) Wheel hoe
       Wheels are provided on these tools for facilitating the weeding operation. Wheel becomes useful in guiding the tool and maintaining the working depth in the soil.
   (e) Manual grubber weeder
       Three tyne hand hoe is known as manual grubber weeder. (See figure)
   (f) Twin wheel weeder
       Two wheels are provided on the weeding tools (as seen in figure) to facilitate and guide the weeding operation.
   (g) Peg-tooth type weeder
       Pegs are provided on surface of the wheel (See figure).
   (h) Roatery paddy weeder
       Roatery paddy weeder is important equipment for interculture operation in paddy cultivation. It is used for uprooting weeds and burying them in puddle soil between rows of standing paddy crop. It improves aeration in soil. It consists of frame, weeding roll, tines, float and handle.
   (i) Manual cono-weeder
       Two cones are fitted adjacent to each other and angled in opposite for better weeding performance in the paddy fields.

   Animal drawn weeding implements
   (a) Animal drawn blade harrow

It is a most common type of implement used for harrowing and weeding operations. It is also known as bakhar. In clayey soils it is used for preparing seedbeds. The primary function of this implement is to pulverize and invert the soil and create the appropriate soil mulch. Mulching is performed to provide covering on soil surface to reduce the evaporation of soil moisture.

(b) Animal drawn improved bakhar

Animal drawn improved bakhar is shown in the figure. The roller wheel provided in rear assists in leveling of the seedbed.

(c) Animal drawn three tyne cultivator, Bardoli hoe, animal drawn two row type blade hoe are some of the other animal drawn weeding implements used widely by Indian farmers.

Tractor operated weeding implements

If tractor drawn weeding implements are to be used, wider spacing for movement of tractors or mini-tractors is necessary. These implements are used after the crop has grown up to some heights to remove the weeds. The main soil working tools i.e. sweeps, shovels etc. are made of high carbon or spring steel. These implements are employed for inter-row cultivation to remove the weeds in the fields with standing crops. Some of them are:

(a) Tractor drawn blade harrow
(b) Tractor drawn two row type blade harrow
(c) Cultivators
(d) Duck foot cultivator etc.

Power weeders

Weeders with their own power units are called power weeders. (See figure). Now-a-days, power weeders are becoming available in the market for conducting weeding operations with least labour requirements.

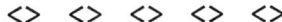

1. INTRODUCTION TO FARM MACHINERY AND EQUIPMENT

Depending on the field operation required for agricultural crop production activities, various kinds of tools, implements and machines are used on agricultural farms. Individual working-elements such as shovel, disc, blade etc. may be known as Tools. An equipment used for conducting an agricultural field operation is generally known as agricultural implement. Generally, an implement does not have any individual moving parts driven by any in-built source of power. Machine is constructed by combination of various individual components having their own motions or movement. In other words, machine is a system or mechanism consisting of various moving components having definite motions of their own. Here, motions may be linear, rotary or reciprocating.

The course FMP-201 (entitled as Farm Machinery and Equipment–I) is conceptualized to introduce the second year students of UG programme in Agricultural Engineering to the farm implements and machines mainly used for crop production on the farms. The syllabus accompanied by appropriate practical studies included in practical manual, contains following major topics in theory part:
1. Agricultural mechanization
   Agricultural mechanization here refers to the process of mechanizing the various field operations which are necessary to conduct in the fields for getting agricultural crop production from the soils. It is also more commonly known as farm mechanization.
2. Classification of farm machineries
   Today, on agricultural farms, various types of farm machinery are used for different kind of agricultural field operations. Farm machinery and equipment which are generally in use may be classified on basis of the kind of field operations to be conducted on the agricultural farms. The different kind of common field operations are: (i) tillage operations, (ii) sowing, planting and transplanting operations, (iii) weeding operations, (iv) plant protection, (v) crop harvesting operations, (vi) threshing operation etc.
3. Principles of operation and selection of farm machines
4. Tillage implements
5. Forces acting on tillage tools
6. Sowing, planting and transplanting equipment
7. Hitching systems and controls
8. Draft measurement of tillage implements
9. Weed control and plant protection equipment
10. Earth moving equipments
11. Measurement of field capacity
12. Calculation of economics of conducting different agricultural operations and use of farm machines

Tillage – Primary tillage and secondary tillage
Tillage is the mechanical manipulation of soil to prepare the proper seed bed suitable for germination of seeds and better growth of the plants. Tillage is divided into two categories: primary tillage and secondary tillage.

Primary tillage is conducted with a view to open up the land through cutting and breaking of the soil pan for making the soil suitable for seedbed preparation. **Primary tillage implements** comprise various types of ploughs such as animal drawn walking type or riding type plough, mould-board plough, disc plough, chisel plough etc. Desi or indigenous ploughs are generally soil stirring ploughs. Mould-board ploughs are soil turning ploughs. The M.B. plough contains mould boards for turning of the furrow slice which is cut soil. Main parts of the mould-board plough are:1. share, 2. mould-board, 3. land side, 4. frog, and 5. tail piece. Disc plough is a rolling plough (Desi plough and M.B. Plough are sliding type ploughs). In Disc plough, spherical discs are used as soil working tools. The main frame of the plough consists of separately mounted steel discs of larger sizes, which rotate while operational.

Tractor drawn ploughs may be either trailing type or mounted type. Chisel ploughs are used for deep ploughing. Subsoiling is conducted by sub-surface ploughs called subsoilers for ploughing the soil at a particular depth with no variation or disturbance on soil surface.

Secondary tillage implements comprise harrows, cultivators, levelers, clod crushers etc. Disc harrows may be animal drawn or tractor operated. Disc harrows may be classified as: single action, double action and offset type disc harrows. Other harrows comprise drag harrows and blade harrows.
Some of the harrows are listed at below.

9. Spike tooth harrow
10. Spring tooth harrow
11. Blade harrows
12. Knife harrow
13. Patela
14. Triangular harrow
15. Zig-zag harrow
16. Bakhar, Guntaka etc.
17. Roller harrow, Rotary harrow etc.

Sowing, planting and transplanting

Sowing is the field operation which is conducted to place the seeds into the soil at proper depth for its germination. Sowing is accomplished in several ways:

(i) Broadcasting
(ii) Sowing behind the plough
(iii) Dibbling
(iv) Seed-drilling (by using animal drawn and tractor operated seed drills)
(v) planters
(vi) trans-planters etc.

Seed drills are used for drilling the seeds at uniform rate of application. Row to row distance is maintained by the seed drills. **Planters** are similar equipment which are, in addition to maintaining row to row distance, used for maintaining the uniform distance between plant to plant.

Trans-planters are employed for planting of the seedlings in place of seeds. Seedlings are raised separately into a smaller sized plot of the field.

Weeding (Inter-culturing)

Weeding is the field operation for removal of the weeds from the agricultural land under cultivation. Weeds are those plants which are not useful for any purpose but they grow into the soil and retard the growth of the agricultural crops grown into that field by consuming precious soil resources. Many types of manual, animal drawn and power operated weed control implements are used on agricultural farms.

Plant protection

Plant protection is the field operation which is conducted to spray the pesticides into the agricultural fields to save the crops against pests and diseases which are responsible for partial or complete damage to the plants. Plant protection equipment includes various types of sprayers and dusters.

Harvesting

Harvesting operations are those field operations which are conducted to cut the crop which is ready to harvest from the fields. Harvesting may be conducted manually by using sickles (a very common hand tool used by the farmers) or by several power driven implements or machines. Reapers, mowers, diggers (for the crops like ground nut and potato) and combine harvesters are used for harvesting of different agricultural crops from the fields.

Threshing

Crops harvested from the fields contain vegetative materials known as straw. Generally, threshing is the field operation which is commonly conducted to separate the grains from the straw. Straw generally contains stems and stalks. Threshing also separates the grains from the cobs and pods. It can be performed manually, by animal driven implements or by power driven machines which are called threshers.

Earth moving equipments

Earth moving equipments are used for land development works. Construction and working of different earth moving equipments such as bulldozer, trencher, elevators etc. is to be studied.

Principles of operation and selection of farm machinery and implements

The farmers always need to conduct the various agricultural operations on time. For this purpose, they inevitably require the different farm implements and machines. Use of farm machines also increases the efficiency of the man and animal power on the farm. Numbers of operations are required for agricultural crop production. Among them, field operations dominate. Operation wise many farm implements and machines are conventionally used by the farmers. Development of various manual and animal driven improved tools and implements has also made their way in utilization by the farmers for use. Tractor power on the farm has considerably reduced the human drudgery involved in the different farm operations like tillage, sowing, harvesting etc. Power tillers are also used by farmers, however it is observed that the use of power tillers as compared to tractors is very less. Mini tractors are also making their way in the Indian market by gaining popularity among the farming community having smaller land holdings. Various factors affect the selection of farm machinery and implements on the farm such as major crops and cropping pattern of the region, cost of labours, size of land holding, availability of water for irrigation, etc.

Other topics

Forces acting on tillage tools, hitching systems and controls, draft measurement of tillage implements are also appropriately covered for study in separate chapters. Methods of measuring the **field capacity** of different farm machinery and implements are discussed. Calculation of **economics** of conducting different agricultural operations and use of farm machines is also to be learnt under this course of study.

2. FARM MECHANIZATION

Mechanization in Agriculture

Hand tools, bullock drawn equipments & power driven machines including the prime movers such as tractors / power tillers may be judiciously employed for performing various operations required for crop production activities.

Mechanization in agriculture refers to the use of machines in agriculture to ensure reduction of drudgery associated with various farm operations. It also enhances efficient and economic utilization of inputs and helps in harnessing the soil and water resources to their best.

The farm mechanization process in fact means judicious application of agricultural machinery / equipment for efficient use of the inputs such as soil, water, seeds, fertilizers, pesticides etc.

Objectives of Farm Mechanization

Objectives of farm mechanization are many. Several of them may be summarized as below:

1. To overcome the shortage of agricultural labour arising due to greater demands during peak period of cropping season.
2. To reduce drudgery involved in farming operations which keeps youths away from adopting the farming as their profession or source of livelihood.
3. To maintain timeliness in conducting farm operations. If farm operations such as ploughing, cultivating, sowing, interculturing, harvesting etc. are not conducted at proper time then the risk of losing the whole cropping season arises.
4. To reduce losses at different stages of crop production. If the crop which is ready for harvesting in the field is not harvested timely then the chances of loss of agricultural produce due to natural or other reasons increases.
5. To improve dignity of farmer. Farmers socially feel pride in having the tractor of their own. In fact, they are empowered due to many reasons by owning of a modern farm machine such as tractor as they does not act as a source of power on the fields but many a times they carry the transportation needs of the rural people.

Present status of farm mechanization

The utilization of farm power for operating the agricultural implements and machines in terms of energy consumption per unit area is one of the parameter of knowing the level of mechanization in any country or state. It may normally be expressed as kW/ha. In India, There has been noticed a continuous increase in the farm mechanization. Among all sources of farm power, the tractor has served as an active catalyst in the process of mechanization on the Indian farms. The tractors are being purchased in large numbers by farmers for self use and also for custom hiring. The major share of mechanical power available on Indian farms is consumed by irrigation pumps. Irrigation shared more than 64 % of the total farm power. After passing five decades in post-independence era, the quantum of animal power has reduced from 0.14 to 0.09 kW/ha. But during this time span, the mechanical power has increased from 0.005 to 0.59 kW/ha. Hence the ratio of mechanical power to total farm power has Increased from 3.6 % to the tune of 80 %.

As in 1996, against 185.4 mha gross cropped area of the country, the total farm power availability was 136.79 mkW hence resulting in the unit farm power availability to 0.74 kW/ha. Mechanical power contributed around 80 percent in the total farm power.

Table 2.1: An overview of the availability of mechanical farm power sources in India

Type of mechanical power	Quantity (in numbers)
Tractor	20,80,000
Power tiller	1,18,000
Combine	6,000
Electric pump	1,48,00,000
Diesel pump	59,40,000
Source: Livestock Census Reports, Automobile Association of India, TERI Energy Data Directory,	

The present annual production of major farm machines is shown in the Table 2.2.

Table 2.2: Annual production data of major farm machines/equipment

Name of farm machine	Approximate annual production (in nos.)
Tractors	2,50,000
Power Tillers	10,000
Pumping Sets	10,00,000
Combine harvesters	2000

(Source: Elements of Agricultural Engineering By Dr. J. Sahay 2004)

Tractor power is largely utilized as a main source of mechanical farm power on the Indian farms. The pattern of utilization of tractor power is indicated in the Table 2.3.

Table 2.3: Annual average utilization of tractor power

Type of work	Annual usage (in hours)
Field operations	230 – 280
Transportation and other stationery applications	420 – 520
Total usage	650 – 800

However, it is estimated that only 19 % of the cropped area is cultivated by tractors having command area of around 15 hectares per tractor. Still there exists great scope for tractors in Indian Agriculture.

3. PRINCIPLES OF OPERATION AND SELECTION OF FARM MACHINERY AND EQUIPMENT

Principles of operation and selection of farm machinery for crop production

A variety of farm machines and implements are available in the local markets in all over India which are useful to conduct one or more field operations necessary to carry out the crop production in the fields. Various field operations concerned with crop production activities can be performed by using manually operated, animal drawn or machine operated farm tools, implements and machinery. Mostly on small farm holdings, the animal drawn implements are very common among the farmers who own the animal power. Increasing availability of tractors has made an impact on this trend with farmers opting for the various tractor operated implements and machines to conduct the various field operations leading to the crop production on agricultural farms. Crop production mainly requires following field operations to be conducted timely during cropping season of different agricultural crops.

- Seed bed preparation
- Seeding operation
- Inter-culture operations (conducted for weeding)
- Plant protection
- Harvesting and threshing of crops

These agricultural operations need different forms of power to carry out in the agricultural farms. e.g. to pull the implements, tractive force is required, which may be obtained from animal or tractor power. Similarly, to operate some of the implements like rotavator, the rotative form of power is required, which may be obtained from the power source like tractors. Pulley power or belt power is required to operate the machines like thresher with no power unit. In such case, rotative power may be obtained from the PTO shaft of the tractor. Automotive power is required to meet the transportation needs of the farms, which can be obtained from animal drawn carts or trailers which can be trailed with the help of tractor, mini-tractor or even by a power tiller.

The need of the farmers to timely and efficiently conduct the necessary field operations for crop production in the fields has motivated the Indian farmers to adopt variety of different farm machines and implements mostly driven by tractors. Tractor has notably reduced the human drudgery involved in the tedious farm operations like tillage, sowing etc. Development of various animal driven improved tools and implements has also attracted the farmers' attention. Power tillers and mini tractors are also making their way in the Indian market by gaining popularity among the Indian farming community and are moving in the direction of providing optional power sources on Indian farms.

Selection of farm machinery and equipment

Selection of farm machinery and equipment depends upon many factors which are necessary to consider before going for purchase of the farm machines. Some of them are narrated at below:

1. Major crops and cropping pattern
2. Availability and cost of labours
3. Size of land holding
4. Availability of finance and cost of machines/implements
5. Type of soil
6. Availability of water for irrigation
7. Availability and type of power source for operating the farm machines
8. Availability of spares and service for the specific farm machinery

4. TILLAGE

Tillage

Tillage means tilling of the soil to prepare it for sowing of the seeds. It is the first activity to be taken up in the process of crop production. Tillage may be defined as the mechanical manipulation of soil to create the favourable conditions in the soil for plant growth. Mechanical manipulation of soil is conducted for making the soil loose and friable which may be suitable for sowing of the agricultural crops. Tillage is also known as seed bed preparation.

Tillage is divided into two categories: primary tillage and secondary tillage.

Several objectives of the tillage are stated at below:
1. To break the hard pan of soil
2. To cut the soil and form the furrows
3. To disturb the compact soil layers
4. To increase the porosity of soil
5. To increase moisture holding capacity of soil
6. To produce aeration inside the soil for growth of the bacteria useful for the plants
7. To prepare the required seedbed suitable for crop growth
8. To do *mulching* for conservation of moisture into the soil

Types of tillage

Tillage is practiced since ancient times. Requirement of tillage depends on type of soil, crops to be grown, average amount of precipitation, availability of water for irrigation, tools and implements used etc. Several types of tillage practices have come into existence over a period of time. Some of them are enlisted and briefly explained at below.
1. Minimum tillage
   Minimum tillage, as its name implies, involves the least manipulation of soil to meet the tillage requirements of the field.
2. Strip tillage
   In this method, tillage operations are carried out in strips.
3. Rotary tillage
   Tillage accomplished through rotary action is known as rotary tillage.
4. Mulch tillage
   Plant residues or other materials are left or spread on soil surface as a part of tillage.
5. Deep tillage
   Tillage performed at a greater depth of soil is known as deep tillage.

5. PRIMARY TILLAGE IMPLEMENTS

Tillage
Tillage means tilling of the soil to prepare it for sowing of the seeds. In other words, tillage is nothing but mechanical manipulation of soil conducted for making the soil loose and friable which may be suitable for sowing of the agricultural crops. Tillage is also known as seed bed preparation. Tillage is divided into two categories: primary tillage and secondary tillage.

Primary tillage
Primary tillage is conducted with a view to open up the land through cutting and breaking of the soil pan for making the soil suitable for seedbed preparation. Primary tillage is conducted by various ploughs such as M.B. Plough, Disc Plough, Chiesel Plough etc.

Primary tillage implements   *(e.g. Desi plough, Disc plough, M.B. plough, Subsoiler)*
Ploughs are used as primary tillage implements. Primary tillage is performed to cut, break and open the soil. The field operation conducted by ploughs is also called as ploughing operation. The various types of ploughs are used for conducting the ploughing i.e. primary tillage operation in the fields. On various basis, ploughs are divided into following types.
On basis of functional components, ploughs are:
> (i) rolling type
>> e.g. disc plough
> (ii) sliding type.
>> e.g. mould-board plough

On basis of source of power, ploughs are:
> (i) animal drawn
>> Which may be walking type or riding type
> (ii) tractor drawn
>> Which may be trailing type or mounted type

On basis of depth of ploughing, ploughs are:
> (i) surface ploughs
>> e.g. mould-board plough
> (ii) sub-surface ploughs
>> e.g. subsoiler plough

On basis of functional characteristics, ploughs are:
> (i) soil stirring ploughs
>> e.g. desi plough (indigenous plough)
> (ii) soil turning ploughs
>> e.g. mould board plough

Several important ploughs are discussed at below.

(i) Desi Plough (Indigenous Plough)
The conventional ploughs which are used by the farmers traditionally may be called as Desi Plough. Desi Ploughs are the indigenous ploughs, which are made by the farmers according to their needs and suitability. The shape and size of these ploughs vary depending on the regions, locations and habits of the farmers. Mainly these ploughs consist of following parts:

(i) Main body, (ii) Shoe, (iii) Share, (iv) Beam and (v) Handle

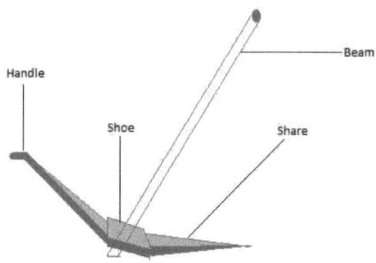

Fig. 4.1: Indigenous (Desi) Plough

(ii)   Mould-board plough

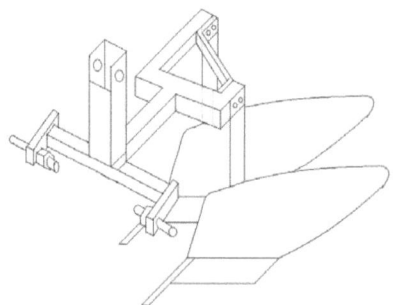

Fig. 4.2: Diagram of a two-bottom M.B. Plough

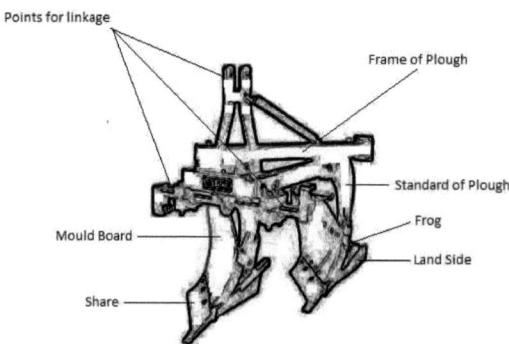

Fig. 4.3: Mould-board Plough

[25]

Main parts of the mould-board plough are:

    1. share, 2. mould-board, 3. land side, 4. frog, and 5. tail piece.

1. Share – it cuts and penetrates into the soil. It is made of chilled cast iron
2. Mould-board – it lifts the soil over it to turn and pulverize
3. Land side –It is made of flat plate. It bears force due to soil thrust
4. Frog – it work as support. All other parts of the plough are attached to it
5. Tail piece – It is a tail part. It facilitates turning of cut soil

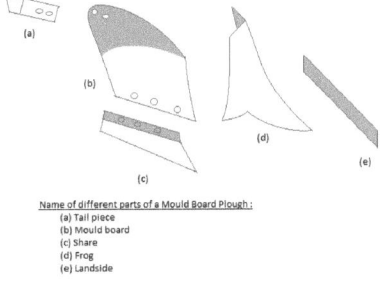

Name of different parts of a Mould Board Plough :
(a) Tail piece
(b) Mould board
(c) Share
(d) Frog
(e) Landside

Fig. 4.4: Different parts of Mould board Plough

Attachments of M.B. Plough are:

1. *Coulter, 2. Gauge Wheel*
1. Coulter – it is attached at ahead of the plough bottom. It makes a vertical cut into the soil for cutting of the furrow slice.
   Rolling type coulter is suspended on shank of the plough bottom.

   Sliding type knife coulter is normally fixed on beam of the plough.

   They help in cutting trashes to facilitate better inversion.

2. Gauge wheel – Gauge wheel is generally provided to maintain the uniform depth of ploughing. It is fitted in hanging position.

Types of share are:

    Slip, slip nose, shin, bar

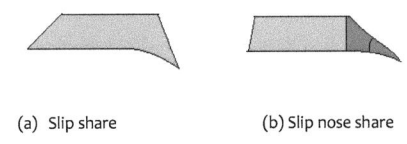

(a) Slip share   (b) Slip nose share

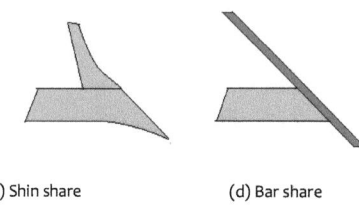

(c) Shin share   (d) Bar share

Fig. 4.5: Different types of share

Types of the mould-board are:

General purpose, stubble, sod, slat

General purpose – General farm use for good pulverization

Stubble – it is relatively short and broad mould board

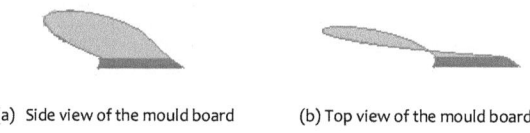

(a) Side view of the mould board   (b) Top view of the mould board

Fig. 4.6: Side and top view of the mould boards

Other parts of M.B. Plough are:

Cleavage end
        The edge of share joining the mould board on frog is known as cleavage end
Vertical clevis
        It is a vertical plate having number of holes on it to adjust the line of pull and control the depth of ploughing
Horizontal clevis
        Horizontal clevis is provided to adjust the plough in horizontal plane with reference to the line of pull to control the width of operation

Several terms related to M.B. Plough are:

Throat clearance
        The vertical distance at the point of share from the beam of the plough
Plough size
        The size of the M.B. Plough is expressed by width of cut, made by the plough, into the soil

Adjustment in M.B. Plough

In Mould Board Ploughs, suction of the plough (horizontal and vertical) is provided to adjust the plough to increase the width or depth of the furrow.

Horizontal suction

It is the maximum clearance (observed from the top) between the land side of the plough and vertical plane touching at the point of share (gunnel side) and heal of land side. Furrow width may be increased or decreased by changing the horizontal suction of the M.B. Plough

Vertical suction

Vertical suction or vertical clearance is the maximum clearance (observed from the side) between land side and horizontal surface. By changing the vertical suction, the depth of furrow can be increased or decreased

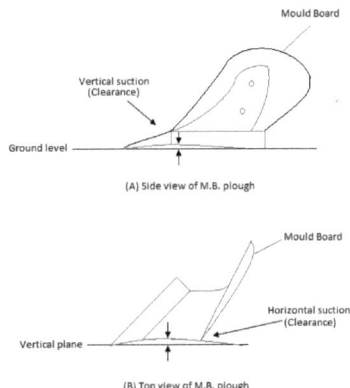

Fig. 4.7: Horizontal and vertical suction in M.B. plough

Reversible M.B. Plough

Reversible M.B. plough consists of two sets of M.B. plough bottoms arranged opposite to each other and capable of rotating to cut and turn the furrow slice towards desired side i.e. left or right.

(iii) Disc plough

The main frame of the disc plough consists of steel discs of larger sizes, which rotate while the plough is made operational. Principally, the disc plough is a rolling plough. Spherical discs (also known as concave discs) are used as soil working tools.

Types of Disc plough

Disc plough may be a *mounted type* or *a trailed type*. Disc plough may be classified into two types:

1. Standard Disc plough

In this plough, the separately mounted discs are set at certain angle to the direction of travel. Diameter of discs vary between 60 to 90 cm. Plough bottom consists of steel disc(s) fitted on stub axle(s) rotating in a taper roller or thrust type bearings carried at the lower end of cast iron supports suspended from the main frame of the plough.

[28]

2. Vertical Disc plough

This kind of disc plough has combined characteristics of a disc plough and a disc harrow. Therefore it is also known as a Harrow plough. In this plough, all discs are mounted on a single shaft which rotates as a single unit like gang of the disc harrow. The size of the discs are smaller as compared to standard disc plough (50 to 65 cm diameter).

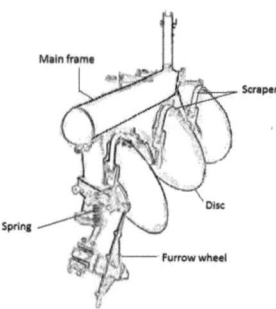

Fig. 4.8: Disc Plough

Comparison of Mould-board plough and Disc plough

Mould-board plough	Disc plough
1. It has a sliding type plough bottom	1. It has a rolling type plough bottom
2. Clogging occurs in wet soil	2. It has less clogging. It is preferable for working in sticky soils
3. Makes use of share to penetrate into the soil and to cut the furrow slice which is lifted and turned by the mould-boards for better pulverization of soil	3. Makes use of discs to cut, break and turn the soil
4. It is not suitable for stony soils	4. It is suitable for working in sticky and stony soils
5. Horizontal and vertical suction are provided for better penetration and efficient working of MB	5. Disc angle and tilt angle are adjusted for better penetration and

plough

Vertical suction helps the plough to penetrate into the soil

Furrow width may be increased or decreased by changing the horizontal suction of the M.B. Plough

The depth of furrow can be increased or decreased by changing the vertical suction

6. Types of mould-board ploughs are:
   a. Animal drawn mould-board ploughs
   b. Tractor drawn mould-board ploughs
   c. Single M.B. plough
   d. Double M.B. or Multi M.B. plough
   e. Trailed type and mounted type
7. Main components of M.B. plough are:
   a. Share
   b. Mould-board
   c. Landside
   d. Frog
   e. Tail piece
   f. Main frame
8. Adjustments in mould-board plough are:
   a. Horizontal suction
   b. Vertical suction

9. Accessories in mould-board plough are:
   a. Coulter
   b. Gauge wheel or furrow wheel

effective operation by disc plough

Increase in disc angle increase the penetration

Reduction in tilt angle increase the penetration

Additional weight on disc plough increase the penetration

6. Types of disc plough
   a. Animal drawn Disc plough
   b. Standard Disc plough
   c. Vertical Disc plough or Harrow plough
   d. Single disc animal drawn plough
   e. Three or four bottom disc plough

7. Main components of Disc plough are:
   a. Discs
   b. Scraper
   c. Plough standard
   d. Bearing housing
   e. Main frame
   f. Adjusting lever
8. Adjustments in Disc plough are:
   a. Disc angle
   b. Tilt angle

9. Accessories in Disc plough are:
   a. Furrow wheel
   b. Spring

Disc angle

The angle made by the plane of cutting edge of the disc with direction of travel is called disc angle. It varies from $42°$ to $45°$.

Tilt angle

The angle made by the plane of cutting edge on ground surface with vertical line is called tilt angle of the disc plough. It varies from $15°$ to $25°$.

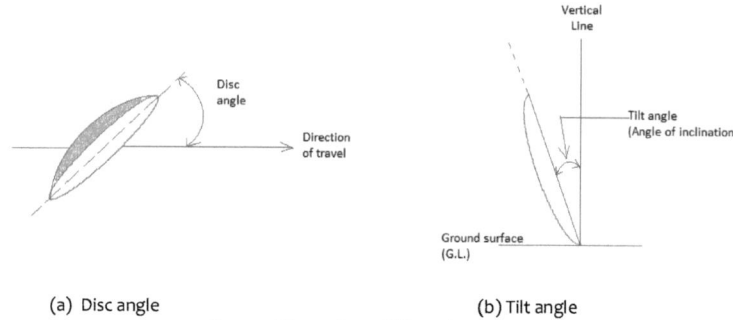

(a) Disc angle   (b) Tilt angle

Fig. 4.9: Disc angle and Tilt angle

(iv)   Chiesel plough

Chiesel plough is used to cut and break the hard soil by using number of narrow tines having chisels as working tools.

(v)   Subsoiler plough

It is a plough which penetrate into the greater depth of soil than the normal deoth of ploughing to break the deeper hard pan of soil with minimum disturbance of the soil surface. It works at the depth of 40-100 cm. It helps to drain excess water in the heavy soils. It helps to conserve moisture into the deeper part of soil in moisture stress conditions. Single tine or two tine subsoilers are commonly used.

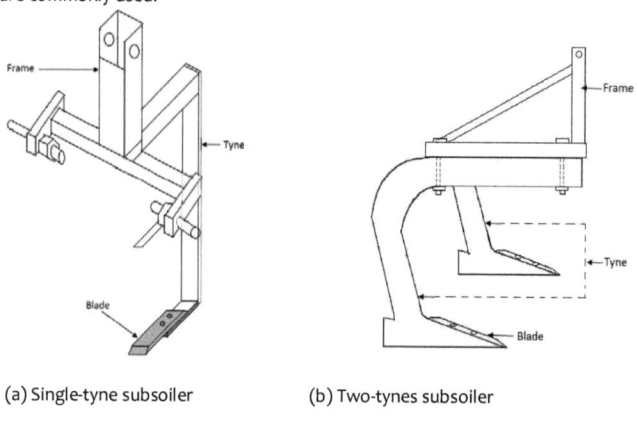

(a) Single-tyne subsoiler   (b) Two-tynes subsoiler

Fig. 4.10: A single-tyne and two-tyne subsoiler ploughs

Subsoiler is designed to perform the ploughing at greater depth than the depth of normal ploughing. Depth of subsoiling normally ranges between 40 to 100 cm. Subsoiler plough consists of a single or multi number of tynes fitted with appropriate blades on it. As standard of

[31]

the plough works beneath the soil surface, upper soil remains undisturbed while soil inside is loosened. Diagram of s subsoiler tyne of a two tyne mini-subsoiler developed by GAU has been shown in the Fig. –.

(vi) **Rotary plough or Rotary tiller or Rotavator**

It is a tillage equipment which has number of blades mounted on a rotating shaft. It is used for seedbed preparation with good weeding and mixing of the crop residue with the soil. Safety guard is provided at rear part behind the rotating blades.

Advantages of rotary tiller
   a. Rapid seed bed preparation
   b. Reduction in draft as compared to conventional tillage implements
   c. In place of using tractive force through tractor wheels, engine power through PTO is utilized to work into the soils

Disadvantage of rotary tiller
   d. power requirement is three times more as compared to power requirement by mould board plough

Principle of operation

Rotary tillers directly utilizes the tractor engine power through PTO, so that slippage of wheel is reduced and excessive weight by tractor in pulling the tillage implement is avoided. Generally three pairs of blades are fitted on each flange clamped to the rotor shaft. As the rotor shaft rotates, the blade cuts a slice from the unploughed soil. Slow rotor speed causes large blade cut and high rotor speed reduces the cut of blade giving fine tilling. High rotor speed requires more power. If rotor speed and engine speed are constant low gear will produce fine tillage while high gear will give rough tillage. If rear shield is lowered, the soils cut by the blades will further be broken to result in levelling effect on soil surface. If rear shield is raised, the soils cut by the blades will not be broken and accordingly power requirement will be reduced. In wet and sticky soils, one pair of blade may be removed for smooth operation but speed of rotor is increased.

Components of rotary tillers

The main components of a rotary tiller are described at below.

Name of component	Description
1. Gear box	Gearbox is useful in reducing the rotating speed received from PTO for getting desired speed of rotation to the rotor shaft. The ratio of speed reduction may be constant or adjustable.
2. Blade or knives	Right hand or left hand blades or knives are mounted on rotor shaft through flanges and arranged in such a way that not more than one blade strikes on soil at a time. 24 to 36 blades are bolted to the flanges connected with the rotor shaft.
3. Side drive	Chain and sprocket assembly or a 3 spur gear assembly is employed for transmitting the drive from the gear box to the rotor shaft.

Safety feature of the rotary tiller

A slip clutch is provided to disengage automatically when safety setting is crossed. Sometimes a shear bolt is used in place of safety clutch.

Size of cut by rotary tiller

The size of cut by a rotary tiller or rotavator is determined by forward speed of the tractor, number of blades and speed of the rotor.

Power transmission in rotavator

Power transmission in rotavator is obtained through employing a shielded double universal joint shaft from PTO to gear box. The gear box hosts a crown wheel & pinion and interchangeable pairs of spur gears. The drive from gearbox is then transmitted to the rotor shaft.

Maintenance of rotavator

Daily maintenance	e. Lubricating and greasing f. Checking oil level in gear box g. Check worn/broken parts h. Tighten bolts
Periodical maintenance	i. Replace bent or damaged blades j. Drain gear box, flush and refill k. Check and lubricate shaft bearings l. Check all bolts m. Check rotor and scroll pattern of blades

6. SECONDARY TILLAGE IMPLEMENTS

Secondary tillage

Primary tillage is followed by the secondary tillage operations with an objective to create proper seedbed through pulverization and inversion of soil to later conduct seeding and planting operations. Secondary tillage is achieved by application of different types of harrows, cultivators, levelers etc. These implements cut the grasses & weeds, mix them into the soil, break the clods & soil aggregates, separate the soil particles and make the soil surface uniform.

Secondary tillage implements

Harrows, cultivators, levelers, clod crushers are used as secondary tillage implements. Secondary tillage implements are used for conducting tillage at shallower depths.

Disc Harrows

On basis of power source, Disc harrows are two types:

   (i)   Animal drawn disc harrows
   (ii)   Tractor operated disc harrows

Disc harrows may also be classified as:

   (i)   Single action
   (ii)   Double action
   (iii)   Tandem
   (iv)   Offset

Single action

   Single action disc harrows operate once on the field throwing the soil in single direction. If it has two gangs, they are placed end to end in such a way that the discs of the right side gang throws the soil towards right and discs of the left side gang throws the soil on left. It cuts the soil in one direction only.

Double action

   Double action disc harrows operates the field twice, each time throwing the soil in two opposite directions. It has one or two gangs at the front and rear. The rear gang follows front gang with the discs mounted in the opposite direction. Gangs are arranged either in tandem or offset.

Tandem

   Tandem disc harrow is made of four gangs in such a way that two gangs are angled at front in opposite directions, which is followed by another two gangs again angled and rotating in opposite directions. (See figure)

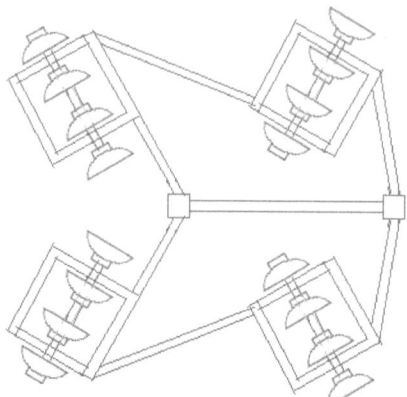

Fig. 6.1: Tandem disc harrow

Offset

Offset disc harrow is a double action disc harrow arranged in offset position on any one side of the line of pull. It is found suitable in orchard crops and gardens. It travels behind the tractor on left or right side of the centre. In case of offset disc harrow, the side thrust occurring due to soil thrust on front gang is balanced by side thrust of the rear gang.

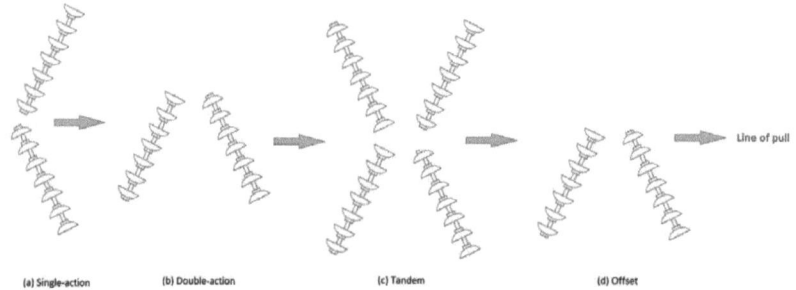

Fig. 6.2: Different types of disc harrow

Components of a disc harrow

1. Disc
2. Gang
3. Gang bolt
4. Gang control lever
5. Spools
6. Bearings
7. Gauge wheel
8. Scraper
9. Main frame etc.

Discs

[35]

Discs are made of high carbon steel having average diameter 35 to 70 cm. plain discs are used for normal works. Discs with serrated edges better cut the vegetative materials. The spacing between discs is kept 25 to 30 cm for heavy duty harrows and less for light duty harrows.

Gang

Gang is made of several discs mounted on a shaft separated with spacers or spools and acts as a composite assembly i.e. all discs rotate at once. The shaft of the gang is also known as arbor bolt, it may be square or round. Discs are tightened with the help of washers and nuts.

Bearings

Bearings are employed to hold the gangs in bearing housings. Due to angle kept among gangs, while moving, bearings are subjected to the axial thrust.

Frame

Frame is the supporting structure of the disc harrow. Angle mechanism and draw bar are also connected with the frame.

Scraper

Scrapers are provided on the discs to prevent clogging.

Depth of harrowing

The depth of harrowing may be increased by:
- Increasing the weight on disc harrow
- Lowering the hitching point
- Using sharp edged discs
- Decreasing speed of operation

Maintenance of disc harrow
- Greasing of bearings
- Tightening of the nuts and bolts
- Sharpening of disc edges
- Replacing worn parts
-

Other harrows

On basis of power source, other harrows may be:
(i)   Animal drawn harrows
(ii)   Tractor operated harrows.

Some of the other harrows are listed at below.
18. Spike tooth harrow
19. Spring tooth harrow
20. Blade harrows
21. Knife harrow
22. Patela
23. Triangular harrow
24. Zig-zag harrow
25. Bakhar, Guntaka etc.
26. Roller harrow, Rotary harrow etc.
27. Clod crusher
28. Roller

Spike tooth harrows

Spike tooth harrows are made of teeth, tooth bars, tooth clamps, guard rails, levers etc. The teeth are made of steel having square, triangular or circular section. Tooth clamps fasten teeth to the teeth bar. Levers regulate the depth of harrowing by setting of teeth. Vertical setting gives maximum depth. Harrow is hitched using draft hooks.

Spring tyne harrows

The spring tyne harrows consists of elliptical/loop or spring teeth (spring like tynes), tooth bars, tooth bar standards, clamps, frame, lever, clevis, draft link, eveners, draw rods etc. The teeth of spring tooth harrows are made of spring steel and circular in shape. The tilling or working end having 2.5 to 3.5 cm width is kept sharp. Sometimes reversible points are used. The point of hitch is raised or lowered by the clevis.

These harrows penetrate deeper than the spike tooth harrows. They are preferred in the soils containing obstructions like stones or roots inside the soils. Spring tynes are bolted staggered on the frame to prevent from clogging.

Blade harrows

Blade harrows (also known as bakhar, kaliyu or guntaka) are very common among Indian farmers. Generally used in clay soils for preparing the required seedbed. It contains a long blade mounted on a conventional cultivator frame. The blade harrow acts like sweeps and cuts the top surface of the soil. It creates a soil mulch. Similar kind of harrows possessing more than one blades (two blades) are also used for interculturing operations.

Acme harrows

Acme harrows consists of curved knives. The front part of the knife crushes the clods and compacts the soil.

Rotary hoe

Rotary hoe consists of two gangs of hoe wheels placed one behind the other to give double action. The wheels are mounted in a staggered manner. The rear gang wheels extend forward between the wheels of the front gang.

Cultivators

Cultivator is however falls in the category of secondary tillage implements due to shallow tillage obtained by it. But in many where and many a times, cultivators are used to open the soil and to prepare seed bed as like that of primary tillage implements.

The cultivators consist of main frame, tynes, shovels etc. Cultivators with shovels are very common. Shovels, made of high carbon steel, are bolted to the tynes. Sweeps are used in place of shovels for conducting shallow cutting operation on soil surface. Two kind of cultivators are in use: 1. Cultivator with rigid tines, and 2. Cultivator with spring loaded tines.

In cultivators with rigid tines, the tines are rigidly mounted on front and rear tool bars of the main frame. Rigid tines of the cultivator do not deflect during operation. While in spring tine cultivators, tines are loaded with springs. When any obstacle is encountered during operation, spring tines swing back and retain the position without damage. In this cultivators, each tine is provided with heavy coil springs. A pair of gauge wheel may be provided for controlling the depth of operation. The cultivators may be equipped with 9, 11 or 13 tines depending on the requirements.

7. FORCES ACTING ON TILLAGE IMPLEMENTS

During tillage operations, at the time of pulling the implement in the soil, various forces act upon the tillage implement. Main forces are known to us include soil resistance, pulling force by the power unit, gravitational force etc. If force 'F' is acting upon the tillage implement, for the purpose of analysis, we shall take three components of the force in three planes as below.

As shown in figure, for a trailed implement

'D' is the draft (horizontal component) of the implement in the direction of pull;

Then D = F cos θ;

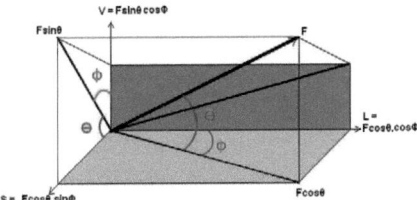

Fig. 7.1: Forces acting on a tillage implement

Here 'θ' is the angle of inclination of force 'F' in the vertical plane it is making with the horizontal plane passing from the center of pull; (see figure)

If 'Φ' is the angle of inclination of force 'F' in the transverse plane it is making with the vertical plane in the direction of travel or draft;

The force 'F' will have 3 components in 3 planes; the longitudinal component 'L' of the draft of the implement by pulling force, the vertical component 'V' and the side force 'S' may be expressed as below:

L = F cos θ. cos Φ   ...   .   (i)

S = F cos θ. sin Φ   ...   .   (ii)

V = F sin θ. cos Φ   ...   .   (iii)

Here, the component 'L' represents the draft of the implement by pulling load, the vertical component 'V' will have the effect of adding load to cause penetration and maintain the working depth of the implement and the side force 'S' will maintain directional stability of the implement while pulled. The side force will have the effect on draft of the implement due to frictional force caused.

In case of mounted implement, the implement will be supported and pulled by tractor; the pulling force of the tractor will be,

$$P \quad = \sqrt{L^2 + V^2}$$

$$= \sqrt{(F\cos\theta.\cos\phi)^2 + (F\sin\theta.\cos\phi)^2}$$

$$= F.\cos \Phi$$

Notes:

1. 'F' is the force exerted by power unit which can have components in all major planes
2. 'θ' is the angle of inclination of force 'F' in the vertical plane with the horizontal surface
3. 'Φ' is the angle of inclination of force 'F' in the transverse plane with the vertical plane

[38]

8. HITCHING SYSTEMS AND CONTROLS

Before going for the field operation implement should be hitched properly with the power source i.e. tractor. The tractors operate at a higher speed as compared to field operations conducted by animal drawn implements. Secondly, to utilize the greater power availability from the tractor, the heavy implements having higher working capacity are chosen. Therefore, for efficient and safe handling of the implement, proper hitching with the tractor is most required.

Hitching methods

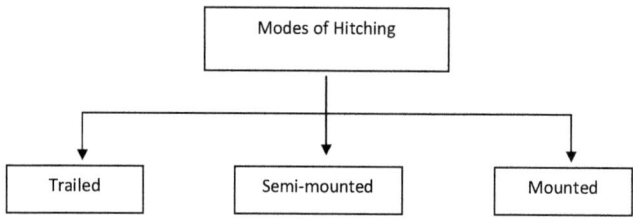

Fig. 8.1: Three methods of hitching

In *trailed* method of hitching the implements, the implement is pulled by a single point of hitch. The weight of the implement is not supported by tractor or any other power source or equipment. e.g. all animal drawn implements are trailed by the animal power. Similarly, tractor drawn trailed disc harrows; trailers etc. are the examples of trailed hitching.

In *semi-mounted* method of hitching of the implements, implement is attached to the tractor and also partly supported by it.

In *mounted* method of hitching of the implements, the implement is fully attached to the tractor with the help of three-point linkage and also hydraulically controlled by the tractor for lowering down or raising up during field operations.

Hitching devices of tractors
While using a tractor for field operation, the implement may be connected or hitched by *drawbar hitch* or *three point linkage* of the tractors.

Drawbar hitch
It is a device or part, which is used to pull the trailing implements. It consists of suitable holes to connect with and fitted on rear of the tractor.

Three point linkage
Three point linkage facility provided at rear of the tractor has been a very convenient arrangement for easy mounting of implements on the tractors. It consists of one upper link and two lower links. These links are operated with hydraulic system of the tractor.

Several important terms related with the hitching of implements are discussed at below:

Centre of power
It is a true hitching point of the power source.

Centre of resistance
It is a point at which resultant force of all horizontal and vertical forces act upon.

Pull
It is the total force required for pulling an implement into the field.

Draft
It is a horizontal component of the pull in the direction of motion.

Line of pull
A line passing through the center of resistance, the clevis and the center of power is called as line of pull.

Side draft
It is the horizontal component of the pull perpendicular to the direction of motion. It causes significance when the center of resistance is not directly behind the center of pull.

Unit draft
The draft per unit cross-sectional area of the furrow is known as unit draft.

9. SOWING EQUIPMENT

Agricultural crop production requires varying farm operations depending on the kind of crops raised on the farm. Sowing or seeding is one of the important farm operations, which is conducted to place the seeds into the soil. Placement of seeds into the soil for the purpose of rearing a crop on the field is known as sowing (seeding) or planting operation. To increase the scope for successful germination of seeds, they should be placed at proper depth and should also be covered by soil. Also, the seeds should be placed at proper spacing (i.e. seed to seed distance). Common methods of sowing in practice are: broadcasting, dibbling, drilling, planting, transplanting etc. In broadcasting, the seeds are thrown on the field with a free hand in scattered manner. Later they are mixed or covered with soil by planking. Seed dropping behind the plough is another method which is common among the farmers. Seed drilling is nowadays become very common method of sowing. Seed drills are used for drilling the seeds into the furrows opened by the furrow openers provided on seed drills. Seeds are also covered with soil simultaneously. Rate of seeds can be regulated in seed drills by employing seed metering devices under the hoppers containing seeds. Different kinds of seed metering devices are used on various seed drills. However, seed to seed spacing in the same row is not uniformly maintained as the seeds are drilled continuously into the soil. Row to row spacing can be controlled by seed drills. Check sowing is conducted in rows perpendicular to each other. In hill dropping, the seeds are dropped at regular spacing. Planters are provided with the arrangement of dropping the seeds at regular interval for maintaining the uniform seed to seed or plant-to-plant spacing in the same row. However the planters are suitable for big sized seeds. In some of the crops, the sowing of seedlings is followed in place of sowing of seeds directly into the soil. e.g. in paddy crop, the seedlings are raised in a separate and small sized plot and then uprooted to plant them in the large field. This method is known as transplanting. Mechanical trans-planters are also developed for conducting the transplanting in labour scarce areas.

Seeding methods

Common seeding methods discussed briefly at above may be summarized as below:
  (i)   Broadcasting
  (ii) Dibbling
  (iii) Drilling
  (iv) Seed dropping behind the plough
  (v)   Hill dropping
  (vi) Check-row sowing
  (vii) Planting
  (viii)   Transplanting

Seed drills

Seed drill is a machine used for placing the seeds in a continuous flow at a uniform rate. Seed drills may be *Animal drawn* or *Tractor drawn.*

Bullock-drawn seed-cum-fertilizer drill

Some machines come equipped with fertilizer attachment to conduct the uniform and simultaneous placement fertilizer also. They are called seed-cum-fertilizer drills, may be drawn by animals or by a tractor. An animal drawn seed-cul-fertilizer drill developed by GAU in 1997 for Bhal and Coastal Agro-climatic region of Gujarat state is shown in Fig. 9.1.

Main parts of the seed-cum-fertilizer drill are: (i) Frame, (ii) Seed box and fertilizer box (hoppers), (iii) seed metering devices, (iv) furrow openers, (v) seed tubes, (vi) Ground wheels, (vii) chain and sprocket mechanism for power transmission, (viii) lever for lifting and lowering of the ground wheel. Transport wheels may be provided for movement of machine from one place to another.

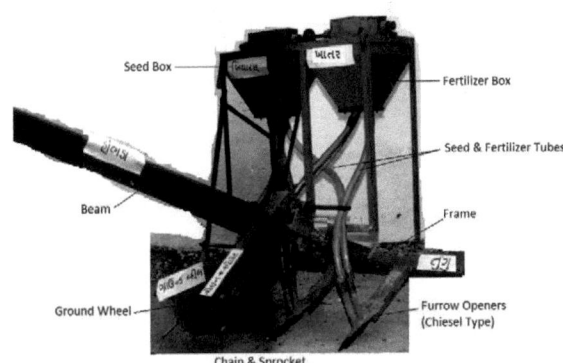

Fig. 9.1: Bullock drawn seed-cum-fertilizer drill developed by GAU for Bhal region of Gujarat

This seed-cum-fertilizer drill was found suitable for sowing of various crops including major kharif and rabi crops of bhal region such as cotton, sorghum, gram, wheat etc. Specifically it was found very much convenient and efficient as compared to local seed drill for sowing of wheat and gram in conserved moisture farming in medium black clayey soils of bhal region of Gujarat. Materials of construction and other specifications are presented in Table 9.1.

Table 9.1: Materials of construction and their specifications for bullock drawn seed-cum-fertilizer drill

Sr. No.	Name of part	Material used and size
1	Beam	G.I. pipe 1" – 2 Nos.
2	Tunga (Frame)	G.I. pipe 2.5" with box sections fitted at different distance
3	Seed and Fertilizer box	M.S. sheet
4	Supporting frame	M.S. angle 1.25" size, 3 mm thickness
5	Driving chain	Rallon ½" pitch
6	Driving gears	M.S. 14 teeth ½" pitch
7	Bevel gear	Mild steel
8	Seed and fertilizer tubes	PVC transparent and flexible pipe
9	Ground wheel	M.S. 3 mm thick plate
10	Shaft	Bright round bar 15 mm dia
11	Furrow openers	1.25" size angles of 3 mm thickness
12	Metering plate	G.I. sheet 18 gauge
13	Nut-Bolts	M.S. as per required size
14	Dadha (Standard)	M.S. square bar
15	Blade	High carbon steel
16	Chiesel of the furrow openers	Gajvel
17	Bushing	Nylon

Tractor drawn seed-cum-fertilizer drill

Tractor drawn seed-cum-fertilizer drills are largely utilized by the farmers. One of such equipment is shown in the Fig. 9.2. It has two separate boxes, one for keeping seeds and another for putting fertilizers with different metering mechanisms. Some equipments are fitted with a large box which is lengthwise partitioned into two compartments, one for seeds and another for fertilizers, both having different types of drilling or dropping arrangements. (Fig. 9.2 (b))

[42]

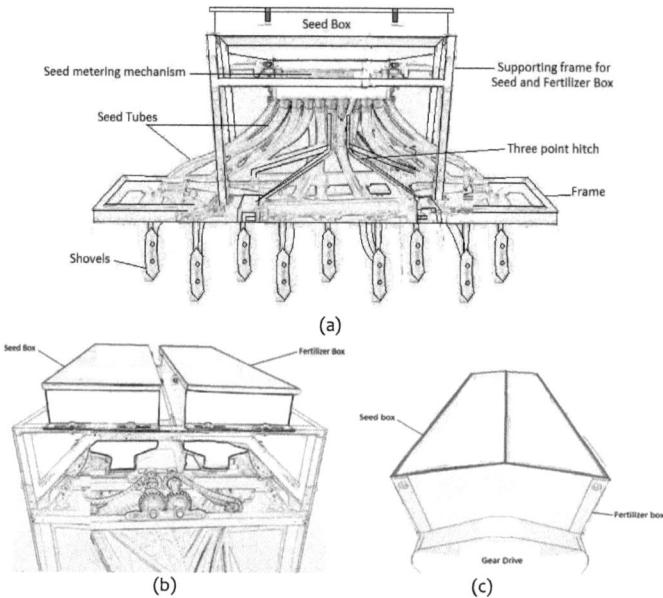

Fig. 9.2: (a) Tractor mounted seed-cum-fertilizer drill
(b) Two separate boxes for seeds and fertilizer
(c) One box divided into two compartments

Tractor drawn seed drills or seed-cum-fertilizer drills are mounted with the help of three-point hitch at rear of the tractors. Number of furrow openers in case of tractor drawn seed drills will be more. Similarly, size of the seed and fertilizer boxes will be greater as compared to animal drawn implement. Functions of a seed-cum-fertilizer drill may be summarized as below:

(i)   It opens furrow at a uniform depth in the soil
(ii)   It meters the seeds at uniform rate and drops in the furrow at uniform depth of soil
(iii)   Fertilizer is also drilled uniformly and placed besides or below the seeds
(iv)   Seeds dropped into the furrow are covered by soil

Seed metering devices
Seed metering or fertilizer metering devices deliver the seeds or fertilizers from the hopper at selected rates. Seed metering devices consists of different kind of metering mechanisms such as following:

(i)   Fluted feed type
(ii)   Internal double run type
(iii)   Cup feed type
(iv)   Cell feed
(v)   Brush feed
(vi)   Auger feed
(vii)   Picker wheel type
(viii)   Star wheel type

[43]

A specific kind of metering mechanism are chosen on basis of the requirements such as size of seeds to be metered, speed of operation, volume or quantity of seeds to be metered etc. Among the list at above, a fluted feed type and internal double run type seed metering mechanisms are explained here.

In *fluted feed* type of seed metering mechanism, equal number of fluted rollers for each furrow opener is provided to carry out metering of the seeds. The roller carries number of grooves on its periphery. The roller rotates with the axle on which it is mounted. Seeds are passed through the seed holes opening into the seed tubes. An adjustable gate is provided at bottom side for discharge of the seeds. The size of opening of the gate is changed according to size of the seeds. Fluted roller has proved a simple, low-cost and trouble-free device mostly suitable for metering of smaller seeds. Spiral shaped flutes are also coming in use. (Note: Bolder seeds may get damaged during metering in fluted rollers)

The *internal double run feed* mechanism for metering contains double faced wheels; one with larger opening suitable for large seeds and another has smaller opening suitable for smaller seeds. Any one of the opening is used at one time while the other not in use is covered by flapper gate placed at bottom of the box or hopper. The speed of the internal feed rollers may be changed by shifting the driving gear on greater or smaller number of teeth provided on the driven gears.

The seed metering devices with orifice and rubber rollers or agitators are also commonly used in tractor drawn seed drills.

Calibration of seed drill
Calibration of seed drill is generally done by the manufacturers to calibrate the seed drill when it is built at first time. Later also, the calibration of seed drills is carried out by the users to confirm the rate of seed metering. As the components of the seed drill wear out over a long usage, the equipment should be recalibrated. Also the calibration of seed drill becomes necessary when seed varieties differ than those in use. It is also always advisable to recalibrate the seed drill before making use of it in the field. Calibration of seed drill is carried out in-house or off the field. The ground wheels are rotated for specific number of revolutions e.g. 100 revolutions to weigh the seeds dropped through the metering devices. By calculating the area covered in 100 revolutions of the ground wheel depending on its width of operation, the seed rate can be calculated in kg per hectare and accordingly proper adjustments may be made to obtain the exact amount of seed rate.

Zero till drill
To avoid extra tillage, for sowing of wheat in the fields of paddy harvested plots, seeding is conducted without conducting any primary or secondary tillage operation. For this purpose, flat furrow openers of narrow width are employed to conduct the seeding operation, and the equipment is known as zero till drill.

Planter
Planters are those seeding machines which are normally used to conduct seeding of larger sized seeds. Seeds, in planters, are not drilled in a continuous flow as it is done in the seed drills, but the seeds are placed one by one. Seed plates are used for this purpose. Planter consists of rotating plates at the bottom of seed hoppers or boxes. In some planters, vertical or inclined rotors are also used. The seed plates move under the arrangement which removes the excess number of seeds except those accommodated in the cells of feed mechanism. Knock out mechanism is also employed to push the seeds placed in cells for dropping in the seed placement device. Usually planters are used for those crops in which the seeds are to be sown at regular intervals in the same row.

Rice transplanters
Rice transplanters are used to transplant the paddy seedlings grown in the nursery in the fields. Transplanters may be manually operated or self propelled one. Manually operated transplanters consists of fixed opening type finger bar which may be actuated by a lever. Normally 2 to 5 seedlings are transplanted in one hill. It saves labour, time and cost of transplanting which is otherwise done

manually. Self propelled type rice transplanters consists of fixed fork and knockout lever type planting fingers. Spacing between hills remain adjustable. Normally one hill contains 3 to 8 seedlings. The machine is built on a float, which prevent its sinking in the field and also found helpful when crossing over the field bunds. It is provided with an engine of around 3 hp for its operation.

Rice seeders
This equipment is used for sowing of pre-germinated paddy seeds in a puddled field by broadcasting in the labour scarce areas where transplanting operation is too much delayed due to absence of labours during peak periods of transplanting. A lugged wheel is employed to transmit the drive to the agitator in the drums. Pre-germinated seeds are kept in drums. Peripheral openings on two sides discharge the seeds. To end the operation, openings are closed.

Tractor mounted sugarcane cutter planter
It is made to receive the whole cane and cuts the sugarcane setts and plants the setts. Mainly it consists of cutting unit, feeding unit and fertilizer metering unit.

Tractor mounted pneumatic planter
Pneumatic planters are used to conduct the precision planting by planting of seeds at predetermined inter and intra-row spacing. A rotating plate having holes on it radially is used. Seed coming in contact with the hole gets stuck on the hole due to suction force and falls down when suction is cutoff at the lowest position near the ground surface. Such precision planters are useful for sowing of very costly seeds and to maintain fixed spacing among the plants and no thinning operation is required.

Problem and Solution
Problem 9.1
The following observations were recorded at the time of calibrating a seed drill.
    (a) Number of furrow openers, n = 6
    (b) Spacing between two furrow openers, d = 20 cm i.e. 0.20 m
    (c) Diameter of drive wheel, D = 0.40 m

Calculate the number of revolutions required to cover the 400 m^2 area.

Solution:

Width of seed drill, w   = n x d
                         = 6 x 0.20 m
                         = 1.20 m
Length of strip required covering 400 sq m area = 400/1.20   = 333.33 m
Circumference of driving wheel   = π D
                                        = 3.14 x 0.40 m
                                        = 1.256 m
Number of revolutions required to cover the area equal to 400 sq m may be calculated as below.
Number of revolutions (N)   = Length of strip / Circumference of drive wheel
                              = 333.33 / 1.256
                              = 265.39 ≈ 265.4 or 265
Allowing about 10 % slippage of the ground wheel during operation, the number of revolutions required to cover 100 sq m area will be
                    265.4 – 0.10 x 265   = 265.4 – 26.5
                                           = 238.86 ≈ 239 revolutions   ...   ... **Ans.**

10. PLANT PROTECTION EQUIPMENT

Farm equipments developed for application of insecticides and pesticides for the purpose of protecting the agricultural crops against the insects, pests and diseases are commonly known as plant protection equipments. Sprayers and dusters are the most common equipments, which are largely used by the farmers for application of liquid and dust formulations respectively. The plant protection equipments may be classified as
- (a) Manually operated, and
- (b) Power operated.

Sprayers

Sprayers are the devices to apply the liquid formulations by spraying action with the help of nozzles to spray the liquid in an atomized form. Depending on the volume of spray per unit area, the sprayers are classified as high-volume, low-volume and ultra-low volume sprayers which are described at below:

High volume spray : Spraying volume is more than 400 litres/ha.
Low volume spray : It ranges from 5 to 400 litres/ha.
Ultra-low volume spray : Spraying volume is less than 5 litres/ha.
Foam spraying : A foaming agent is added to the spraying solution. The special nozzle is used for foam spraying.

Essentially all sprayers work on the principle of operation through developing air compression inside a chamber. Pumping action is employed to generate the air compression. Depending upon the construction and mode of functioning, different kind of sprayers are available such as shown at below.
1. Knapsack sprayer
2. Hand compression sprayer
3. Foot sprayer
4. Stirrup pump sprayer or Bucket sprayer
5. Rocking sprayer

1. Knapsack sprayer

This is one of the most widely used sprayers, which are carried on back of the operator. The knapsack sprayer is fastened with the help of shoulder straps. On one hand (left hand) operating lever is provided to maintain the air compression through pumping, and on another hand (right hand) the spraying lance is provided to operate the spraying. This kind of sprayer consists of a tank for storing the liquid insecticides/pesticides. Reciprocating pump is utilized to build and maintain the compression into the air chamber by a lever provided to operate by the hand. With the help of operating lever, pumping is performed and the air compression is created to enforce the liquid pesticide into spraying lance through delivery tubes. The liquid under pressure is passed through the nozzle to form the minute particles and is sprayed outside in the pattern as desired depending on the type and setting of the nozzle used. Main parts of the knapsack sprayer are:

(i) Tank	(ii) Pressure chamber	(iii) Delivery tube
(iv) Agitator	(v) Delivery valve assembly	(vi) Operating lever
(vii) Delivery hose	(viii) Shut-off cock or cut-off valve	(ix) Spraying lance
(x) Nozzle	(xi) Filling hole cap	(xii) Strainer

2. Hand compression sprayer

The pressure of the compressed air is used to compel the liquid to pass through the delivery hose to the nozzle fitted at the end of a spraying lance. Parts of the hand compression sprayer are:

(i) Tank	(ii) Pumping chamber	(iii) Plunger rod

(iv) Delivery tube   (v) Valve assembly   (vi) Plunger bucket assembly
(vii) Delivery hose   (viii) Shut-off cock or cut-off valve   (ix) Spraying lance
(x) Nozzle   (xi) Filling hole cap

3. Foot sprayer

In foot sprayer, a pedal provided to operate the pumping assembly. A pressure vessel, stand, spring, pedal, suction hose, delivery hose, strainer, shut-off cock/pistol, spraying lance, nozzle are some of the main components of the foot sprayer.

4. Stirrup or bucket sprayer

It consists of a handle, spring, pumping barrel, bucket assembly, suction valve assembly, adjustable stirrup with footrest, spraying lance with nozzle and others. The pump is operated by pressing the footrest with one foot and the plunger is moved up and down to create the sufficient pressure for spraying.

5. Rocking sprayer

It comprises a pressure vessel attached with a pumping assembly, which is operated by a hand lever (a long rod) provided for to and fro movement to create the pressure inside the pressure vessel for spraying. Trigger cut-off valve, spraying lance and nozzle are provided as usual. This sprayer is suitable for tall crops such as trees and many horticultural crops.

Power sprayer

The sprayers equipped with their own power source for its operation are known as power sprayers. Power sources for the sprayer mostly include the power available from a pump operated by an engine or a tractor. Also, due to availability of more power as compared to human power, the power sprayers possess the higher capacity in terms of more coverage of area through operating many nozzles at a time. Nozzles are mounted on booms. The most common parts of a power sprayer are: tank with agitator, frame for mounting of the spraying nozzles, regulator, pressure gauge, etc. A suitable kind of pump is required for creating sufficient pressure to accomplish proper spraying.

Parts of sprayer and their functions

Main parts or components, which play major role in the functioning of the sprayers are explained in brief at below.

1. Nozzle

   Nozzle is one of the most important parts of the sprayer. In the nozzle, the spray liquid is forced under pressure through a small hole or orifice. Various types of nozzles are employed for use in the sprayers depending upon the type of spray required. Common types of nozzle are: hollow cone nozzle, solid cone nozzle and fan type.

2. Swirl plate

   Swirl plate is a part of nozzle, which imparts rotational motion to the liquid particles passing through it.

3. Spraying lance

   It is a hand held pipe. One end of the spray lance is connected with the delivery hose through a cut off valve and another end is mounted with a nozzle.

4. Cut-off valve

   It is provided to control the flow of liquid.

5. Air chamber

   An air chamber is required to maintain the pressure.

6. Agitator

   To agitate the liquid pesticides kept into the tank to maintain uniform density throughout the spraying process.

7. Tank

   Steel or plastic tanks are used to store the liquid to be sprayed.

8. Strainer

Strainer is used to prevent the dust particles and other foreign materials from entering into the sprayer.

9. Boom

It is generally provided on power sprayers on which the more number of nozzles are mounted to operate simultaneously. Nozzles are spaced at the distance suitable for the respective crops. Booms may be vertically adjusted to suit the height of the crops to be sprayed.

Types of spray

According to volume of the spray, the sprayers may be classified into following categories:

1. High volume spray   (more than 400 litres/ha)
2. Low volume spray   (5 to 400 litres/ha)
3. Ultra low volume spray   (less than 5 litres/ha)
4. Foam spraying   (Foaming agent is added)

Duster

Several chemicals used for plant protection measures are available in powder form. Wettable (i.e. soluble) powder may be mixed with water and may be sprayed using an appropriate sprayer. But in case of those, which needs to be applied in dry form needs dusters. Dusters are the equipments, which are constructed to apply the chemicals in solid dust forms. Dusters employ the use of airflow to pass through the chemical powder and carry the dry fine particles of chemical to spread on the affected plants.

Parts of Duster

A duster consists of: (i) Container, (ii) Agitator, (iii) Operating handle, (iv) Fan or Blower, (v) Delivery hose/pipe, (vi) Power source unit (in case of power duster) etc.

Types of Duster

Several types of dusters in use are: (i) Plunger type duster, (ii) Rotary type duster, (iii) Knapsack type duster, and (iv) Power duster.

(i) Plunger type duster

It has a smaller piston which forces the air stream to pass through the dust particles and carry with to pass from the delivery spout for its application on the affected plants.

(ii) Rotary duster

It consists of a hand-operated rotor. The flow of air is produced by a rotary fan which is blown out with the dust (i.e. chemical powder) through a delivery tube. It is used for dusting (i.e. spraying as in case of liquid chemical) tall crops. The rate of flow may be reduced or increased by regulating the fan speed.

(iii) Knapsack type duster

It has a container of powder (dust) chemical, which is carried on back of the operator. The fine particles of dry chemical are blown to the plants for application.

(iv) Power duster

It has a power unit attached which operates the fan to create the stream of air to blow off the dust from the container to spread the dust forcibly at a long distance to fall and spread on the affected plants. The delivery spout is adjustable to regulate the distance and direction of application of chemical dust powder.

Apart from the sprayers and dusters, several customized pesticide applicators are also available which are briefly described here.

Power sprayer cum duster

It is equipped with a suitable power unit to operate the equipment either to act as sprayer or duster depending upon the requirement.

Fumigator

Fumigators are the equipments, which are used to apply the chemicals (i.e. pesticides/insecticides) in gaseous form.

Seed dresser

It is used to apply the protective coating of the appropriate chemicals on the seeds before sowing to ensure safety of seeds before germination.

Suction or discharge capacity of sprayer/duster

As pumps are employed in the sprayers and dusters, the suction or discharge capacity may be calculated by using the following equation:

$$Q = \frac{\pi}{4}.D^2.L.n.10^{-5}$$

Where, Q = suction or discharge capacity in l/min
D = diameter of the plunger in cm
L = length of stroke in cm
n = number of revolutions per minute

Problems and solution

Problem 9.1

Calculate the power required to discharge liquid at the rate of 30 l/min at the pressure of 30 kg/cm^2.

Solution

Given:

Rate of discharge i.e. volume of liquid per unit time (volume in litres and time in minutes)

Considering 1000 litre (l) = 1 m^3
1000 l = 1 m x 1m x 1m
= 100 cm x 100 cm x 100 cm
= 10^6 cm^3

∴   1 l $= \dfrac{10^6}{1000}$ cm^3

= 1000 cm^3

∴   30 l = 30,000 cm^3

$$
\begin{aligned}
\text{Power required} \;&=\; \frac{\text{Work}}{\text{Time}} \\[4pt]
&=\; \frac{\text{Force x Length}}{\text{Time}} \\[4pt]
&=\; \frac{\dfrac{\text{Force}}{\text{Area}} \times (\text{Length x Area})}{\text{Time}} \\[4pt]
&=\; \frac{\text{Force}}{\text{Area}} x \frac{(\text{Length x Area})}{\text{Time}} \\[4pt]
&=\; \text{Pressure x} \; \frac{\text{Volume}}{\text{Time}} \\[4pt]
&=\; \text{Pressure x Discharge} \\[2pt]
&=\; 30 \,\text{kg/cm}^2 \text{ x } 30000 \,\text{cm}^3/\text{min} \\[2pt]
&=\; 30 \text{ x } 30000 \,\text{kg.cm/min} \\[4pt]
&=\; 30 \text{ x } 30000 \text{ x } 9.8 \text{ x } \frac{1}{100} x \frac{1}{60} \text{N.m/s} \\[4pt]
&=\; \frac{30 \text{ x } 30000 \text{ x } 9.8}{100 \text{x} 60} \quad (\text{in N.m/s i.e. J/s} \qquad \text{i.e. watt})
\end{aligned}
$$

$$= 1470 \text{ watt}$$
$$= 1.47 \text{ kW} \quad \ldots \quad \ldots\textbf{Ans.}$$

Problem 9.2
Find the suction capacity of a power sprayer if diameter is 25 mm, speed is 1100 rev/min, length of stroke is 22 mm and number of plunger is 3.

Solution

Diameter, D = 25 mm
Speed, s = 1100 rev/min
Length of stroke, L = 22 mm
Number of plunger, n = 3

The suction capacity is given by $Q = \dfrac{\pi}{4}.D^2.L.n.10^{-5}$

Where,

Q = suction capacity in l/min
D = diameter of the plunger in cm
L = length of stroke in cm
n = number of revolutions per minute

Here,

D	= 25 mm	
	= 2.5 cm	
L	= 22 mm	
	= 2.2 cm	
n	= 1100 revolutions per minute	

$\therefore \quad$ Q $= \dfrac{\pi}{4}.D^2.L.n.10^{-5}$

$= \dfrac{\pi}{4}.(2.5)^2.(2.2).(1\,100).10^{-5}$

$= 0.1187 \text{ l/min} \quad \ldots \quad \ldots \quad \ldots\textbf{Ans.}$

Selection of Plant Protection Equipment
The correct choice of the proper plant protection equipment is required for their economical and effective usage. The major choices are: (i) Dusters, (ii) High-volume sprayers, and (iii) Low-volume sprayers. Selection of a specific spraying / dusting equipment may be made by considering the conditions and requirements.

Conditions and requirements	Recommended equipment
Where water is very scarce and powdery form of pesticides is available...	The dusters are preferred over sprayers to avoid greater usage of the water, particularly in arid and semi-arid regions.
If sprayer is to be chosen for spraying liquid pesticides....	Adjustable sprayer capable of spraying fluid in high volume and low volume both is preferred.
For spraying on small fruit trees...	Foot sprayer and rocking arm sprayers are preferred
For spraying field crops...	Knapsack sprayer is a good option
Large areas having tall and row crops...	Tractor operated or power sprayers
If ultra-low volume spraying is required	Ultra low-volume sprayers

11. EARTH MOVING EQUIPMENT

Equipment for conducting various earth moving operations such as excavation, embankment, elevating, trenching etc. can be known as earth moving equipment. Some of the earth moving operations may also be conducted with the help of suitable attachments mounted on the tractor. Some of the common earth moving operations are described in brief at below.

Earth moving operations
  (i)   Excavation
  (ii) Embankment
  (iii) Ditching
  (iv) Compaction
  (v) Road surfacing
  (vi) Trenching
  (vii)Spreading a stock pile
  (viii)   Back filling
  (ix) Loading
  (x) Material handling
        etc.

Some of the earth moving equipment or attachments are described at below. Some of them are tracked equipment and rest are wheeled ones.
  (i)   Dozers
  (ii) Loaders
  (iii) Excavator
  (iv) Trencher
  (v) Bull dozer
  (vi) Backhoe loaders
  (vii)Rock breaker etc.

Dozer attachment for tractor
The dozer mounted in front of the tractor consists of a thick curved plate with a hardened strip having sharp cutting edge fastened on the dozer plate at the bottom. The strip can be replaced if it has worn or become blunt. The dozer plate mounted on the tractor consists sturdy arms for raising or lowering by hydraulic system of the tractor.

Backhoe attachment
Backhoe attached to the rear of the tractor consists of a bucket with digging fingers. It also consists of hydraulic cylinder and arms. The bucket position is controlled by hydraulic system. The digging fingers are hardened for greater strength against wear and tear and can be replaced if worn or become blunt.

Dozer / Bulldozer
Dozers are basically the crawlers or wheel tractors of higher capacity which is equipped with a heavy curved blade mounted at front end for conducting the earth moving operations. Other functions of the bulldozer/dozer are: cleaning of the ground, road surfacing, leveling etc. It is a powerful machine to conduct the operations which require high drawbar pull and higher traction to perform the functions of land clearing and dozing for pushing a large amount of earth or rock materials used in land development activities for the purpose of construction, road-building, farming and other useful works. Dozer consists of a heavy blade or plate made from steel and mounted on a 4-wheel drive tractor or a crawler type tractor. The blade is lifted up or lowered down with the help of hydraulic rams provided.

Types of blades used on dozers

Blade used on dozer may be straight, angular or special purpose depending upon the kind of works on hand. The constructional and functional features of different blades used on dozers are presented in the Table –.

Table –: Different type of blades used for mounting on dozers

Type of blade	Constructional feature	Functional feature
Straight blade	The blade is mounted in fixed position and perpendicular to the direction of travel Forward or backward tilting and pitching of the blade is allowed upto 10°	used to push the heavier materials and even to cut the ditches
Angular blade	- As above -	Used to make the cuts on hills for the works such as road cuts on hill sides
Special purpose	- As above -	Suitable for clearing trees and vegetations

Fig. 11.1: Tractor mounted with a Dozer Blade

Adjustments for conducting various field operations by using dozers are shown in the Table presented at below:

Table 11.1: Types of field operations and respective settings of the Dozer blade

Field operation required	Adjustment, setting and working of Dozer blade
To conduct digging into the field	The blade of the dozer is maintained at below ground level during field operation
To transport the material from one place to another	The blade is placed exactly at the surface/ground level
To spread the material on to the field	The blade of the dozer is held higher than the ground level
To cut V-ditches of shallower depth	The dozer blade is tilted to cut the shallow V-ditches
To create a stock-pile	Dozer can be used to collect the materials at one place for further loading in the transporting vehicles
To spread the stock-pile	The $1/3^{rd}$ of the blade width can be used to cut the materials from the sides of the stockpile
Backfilling	The back-filling of the ditches can be made by pushing the ditch or foreign materials with the help of dozer

	blade. For this purpose, the blade is kept at angle with the ditch while travelling forward.

Loader / Excavator

The equipment used for loading and unloading of the materials to be displaced is known as loader. The loader is a very versatile kind of equipment which can also be employed to excavate at various heights with the help of suitable shovel or bucket attachment. It may be a self propelled machine fitted with a shovel or bucket at the end of articulated arms.

A bucket made of heavy duty steel and fitted with replaceable cutting edges and teeth is used as attachment to perform the excavation. The multi-purpose bucket can also be employed for use as dozers. Excavators are used for removing the big quantity of soils from the earth.

Some examples are: backhoe loader, compact type excavator, tracked excavator, wheeled excavator etc.

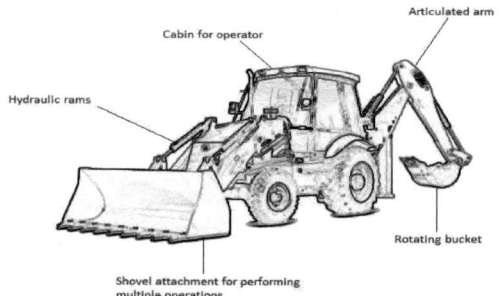

Fig. 11.2: Wheeled backhoe-loader equipment with front and rear attachments

Forklift

A forklift attachment made of steel and provided with two movable tynes is mounted on a suitable tractor unit to act as forklift equipment for material handling purposes.

Trencher

Equipment used for preparing a long narrow ditch is called a trencher. Embankment is also made using the soil obtained from the trenching operation.

Grain elevator

It is an equipment which comprises a belt with scoops for raising the grains to store at upper sides.

12. ECONOMICS OF USING FARM MACHINERY AND EQUIPMENT

Cost Analysis of Farm Machinery

Cost analysis should be worked out to calculate the hourly operating cost of the farm machines and equipments. That may be derived from the summation of fixed and variable costs.

Fixed cost includes:
- i) Depreciation
- ii) Interest on investment
- iii) Taxes & Insurance
- iv) Shelter

Variable cost comprises:
- i) Repairs & Maintenance costs
- ii) Fuels & Lubricants
- iii) Labour costs

Calculation of Depreciation

Depreciation is the reduction in the value of machine over a specific period of time. Machines depreciate due to its oldness, prolonged use, wear and tear. Depreciation is a part of accounting process and reflects the amount of decline in value. Several methods of calculating depreciation are:

1. Estimated value method
2. Compound interest method
3. Constant percentage method
4. Straight-line method

Interest on investment

In the agricultural machinery management, interest is the second largest item of the expenses. It is assumed as the amount of interest on the capital, which would be borrowed to purchase that machine. This may range between 10 to 18 %.

Since the depreciation has been considered regularly over a period of possession, the annual interest is not calculated on basis of full cost of purchase. But it is calculated on the average value (mean value of purchase cost and salvage value).

Therefore, If 'I' is the annual interest, 'P' is the purchase price, 'S' is the salvage value and 'i' is the rate of interest (%) as may be prevailing in the market during the year, then

$$I = \frac{P+S}{2} x \frac{i}{100}$$

Taxes, Insurance & Shelter

Taxes & insurance charges together are taken as 2 % of the purchase price. Cost on shelter is taken as 1 % of 'P'.

Repairs & Maintenance costs

To maintain a machine in a good working condition for ensuring its reliability in the fields, periodic maintenance and occasional repairs are always required. Normally, repair requirements increase after long field use of the machine. In case of tractors, a study showed that the repair and maintenance costs are directly related with the age and inversely related with the annual use. Size in terms of horsepower also affects the repairs & maintenance costs. However for the sake of convenience in estimating the annual operating costs the average repairs & maintenance costs may be worked out on basis of average expenses made during previous years. It can be assumed in terms of certain percent of the purchase price of the machine (it ranges from 8 to 10 % of the cost of purchase in case of agricultural machines).

Costs of Fuel and Lubricants
Fuel costs may be estimated on basis of rate of fuel consumption and estimated use of the machine. Load conditions and size of the engine used in machine are some of the main factors which affect the fuel consumption. Test results may provide more specific information on average fuel consumption. Fuel costs are calculated from the annual machine usage hours and the specific fuel consumption (in litres/hr). Prevailing market rates of the fuel cost is taken into consideration. Cost of lubricants may be taken as around $1/5^{th}$ of the fuel charges.

Labour costs
Manpower will always be required to operate any agricultural machine and therefore as per necessity (depending on type of machine and nature of farm operation) the costs of skilled and unskilled manpower should be taken into consideration.

Example
Determine the hourly cost of operating a 35 hp tractor costing 3,75,000 having useful working life of 12 years with an annual use of approximately 1000 hours. It consumes 4.2 litres of fuel at $3/4^{th}$ load. Driver is paid Rs. 200 per day.

Solution:
To calculate the hourly cost of operation, fixed and variable costs have to be worked out.

(a)   Fixed cost

    (i)   Annual depreciation (in Rupees)

$$D = \frac{P-S}{L}$$

$$= \frac{(375000-37500)}{10}$$

$$= \frac{337500}{10}$$

$$= 33{,}750$$

    (ii)   Annual interest on investment (in Rupees)

$$I = \frac{P+S}{2} \times \frac{i}{100}$$

$$= \frac{375000+37500}{2} \times \frac{12}{100}$$

$$= \frac{412500}{2} \times \frac{12}{100}$$

$$= 24{,}750$$

    (iii)   Taxes & Insurance (in Rupees)

         @ 2 % =   0.02 x 375000

         =   7500

(iv)   Shelter (in Rupees)

@ 1 % =   0.01 x 375000

=   3750

Total fixed cost = 33,750 + 24,750 + 7500 + 3750 = 69,750

(b)   Variable cost

(i)   Repair and Maintenance cost of implements

@ 8 % =   0.08 x 375000

=   **30,000**

(ii)   Cost of fuel

Considering the rate of diesel fuel as Rs. 47 per litre and annual usage of tractor is 1000 hours

∴ Cost of fuel = 47 x 4.2 x1000

= **1,97,400**

(iii)   Cost of lubricants

Assuming the cost of lubricants as $1/4^{th}$ of fuel cost

Cost of lubricants $= \dfrac{1}{4}$ x 197400

= **49350**

(iv)   Cost of driver

Cost of driver per day = 200

Considering working hour per day as 8 (eight),

No. of working days   = 1000 / 8

= 125 days

Cost of driver for 125 days   = 125 x 200   = 25,000

(v)   Hence annual cost of driver   = **25,000**

Total variable cost = 30000 + 197400 + 49350 + 25000 = 3,01,750

Now, total cost of operation by bullocks   = Fixed cost + Variable cost

= 69,750 + 3,01,750

= 3,71,500

Hourly cost of operation on basis of annual use   = 371500 / 1000

= 371.50

∴   Operating cost of tractor per hour   = **371.50**...   ...   **Ans.**

LIST OF REFERENCES

Ali, I., 2011. Farm Power Machinery and Surveying – A Text Book for Degree Classes. Kitab Mahal Publishers, New Delhi.

Anonymous, 1998. Form B. Agricultural Engineering Research Subcommittee Report for the year 1997-98 Presented at GAU, Junagadh.

Jain, S.C. and Philip, G., 2003. Farm Machinery – An Approach. Standard Publishers Distributors, Delhi. 254 p.

Sahay, J., 2004. Elements of Agricultural Engineering. Standard Publishers Distributors, Delhi. 461 p.

Shippen, J.M., Ellin, C.R. and Clover, C.H., 1980. Basic Farm Machinery. Pergamon Press Ltd. 288 p.

Singh, S. and Verma, S.R., 2009. Farm Machinery Maintenance and Management. Directorate of Information and Publications of Agriculture, Indian Council of Agricultural Research, New Delhi. 212 p.